How the Leopard Changed Its Spots

THE EVOLUTION OF COMPLEXITY

Brian Goodwin

A TOUCHSTONE BOOK
Published by Simon & Schuster
New York London Toronto
Sydney Tokyo Singapore

TOUCHSTONE
Rockefeller Center
1230 Avenue of the Americas
New York, NY 10020

First Touchstone Edition 1996

TOUCHSTONE and colophon are registered trademarks
of Simon & Schuster Inc.

Manufactured in the United States of America

10   9   8   7   6   5   4   3   2   1

Library of Congress Cataloging-in-Publication Data
Goodwin, Brian C.
    How the leopard changed its spots: the evolution of
complexity / Brian Goodwin.—1st Touchstone ed.
        p.   cm.
    Originally published: New York: C. Scribner's Sons,
© 1994.
    Includes bibliographical references (p.   ) and index.
    1. Evolution (Biology)   2. Morphology.   3. Self-
organizing systems.   I. Title.
[QH366.2.G655   1996]
575—dc20   95-50307
ISBN 0-684-80451-4 (pbk.)

contents

preface

Scientific theories develop out of choices and assumptions that are neither arbitrary nor inevitable. Darwin made particular assumptions about the properties of organisms and their evolution that have led to one of the most successful theories ever to have emerged in science. He accepted that the major phenomenon of life that needs to be accounted for is the adaptation of organisms to their habitats, and he believed that this could be explained in terms of random hereditary variations among the members of a species and natural selection of the better variants over long periods of evolutionary time. This has become the basis for explaining all aspects of life on earth, or elsewhere. No aspect of human life is untouched by Darwin's theory of evolution, modified in various ways to apply to economics and politics, to the explanation of the origins and the significance of art, and even to the history of ideas themselves.

However, all theories carry with them a particular viewpoint, a way of seeing phenomena that produces sharp focus on certain aspects of reality and blurred vision elsewhere. A striking paradox that has emerged from Darwin's way of approaching biological questions is that organisms, which he took to be primary examples of living nature, have faded away to the point where they no longer exist as fundamental and irreducible units of life. Organisms have been replaced by genes and their products as the basic elements of biological reality. This may seem to fly in the face of all common sense, but stranger things have happened in the name of science. What's more, there is no lack of highly persuasive books whose objective is to demonstrate why organisms are not what they seem to be—integrated entities with lives and natures of their own—but complex molecular machines controlled by the genes carried within them, bearers of the historical record of the species to which the organism belongs. Though this is certainly not what he anticipated, this is in fact the sharp focus that has developed from Darwin's assumptions about the nature of life, and there is no denying the remarkable insights that have accompanied this illumination of the molecular level of organisms.

There is always a price to be paid for excessive preoccupation with one aspect of reality. Modern biology has come to occupy an extreme position in the spectrum of the sciences, dominated by historical explanations in terms of the evolutionary adventures of genes and an associated single-level molecular reductionism of gene products. Physics, on the other hand, has developed explanations of different levels of reality, microscopic and macroscopic, in terms of theories appropriate to these levels, such as quantum mechanics for the behavior of microscopic particles (photons, electrons, quarks) and hydrodynamics for the behavior of macroscopic liquids. It is the absence of any theory of organisms as distinctive entities in their own right, with a characteristic type of dynamic order and organization, that has resulted in their disappearance from the basic conceptual structure of modern biology. They have succumbed to the onslaught of an overwhelming molecular reductionism.

Here we face another curious consequence of Darwin's way of looking at life: despite the power of molecular genetics to reveal the hereditary essences of organisms, the large-scale aspects of evolution remain unexplained, including the origin of species. There is "no clear evidence . . . for the gradual emergence of any evolutionary novelty," says Ernst Mayr, one of the most eminent of contemporary evolutionary biologists. New types of organisms simply appear upon the evolutionary scene, persist for various periods of time, and then become extinct. So Darwin's assumption that the tree of life is a consequence of the gradual accumulation of small hereditary differences appears to be without significant support. Some other process is responsible for the emergent properties of life, those distinctive features that separate one group of organisms from another—fishes and amphibians, worms and insects, horsetails and grasses. Clearly something is missing from biology. It appears that Darwin's theory works for the small-scale aspects of evolution: it can explain the variations and the adaptations within species that produce fine-tuning of varieties to different habitats. The large-scale differences of form between types of organism that are the foundation of biological classification systems seem to require another principle than natural selection operating on small variations, some process that gives rise to distinctly different forms of organism. This is the problem of emergent order in evolution, the origins of novel structures in organisms, which has always been one of the primary foci of attention in biology.

It is here that new theories, themselves recently emerged within mathematics and physics, offer significant insights into the origins of biological order and form. Whereas physicists have traditionally dealt with "simple" systems in the sense that they are made up of few *types* of components, and observed macroscopic (large-scale) order is then explained in terms of uniform interactions between these components, biologists deal with systems (cells, organisms) that are hideously complex, with thousands of different types of genes and molecules all interacting in different ways. Or so it seems at the molecular level. However, what is being recognized within the "sciences of complexity," as studies of these highly diverse systems are called, is that there *are* characteristic types of order that emerge from the interactions of many different components. And the reason is not unlike what happens in "simple" physical systems. Despite the extreme diversity of genes and molecules in organisms, their interactions are limited so that distinctive types of order arise, especially in relation to the large-scale aspects of structure or morphology, and the patterns in time that constitute organismic behavior. A particularly striking property of these complex systems is that even chaotic behavior at one level of activity—molecules or cells or organisms—can give rise to distinctive order at the next level—morphology and behavior. This has resulted in one of the primary refrains of complex studies: order emerges out of chaos. The source of large-scale order in biology may therefore be located in a distinctive type of complexity of the living state that is often described in terms of the computational capacity of the inter-acting components rather than their dynamic behavior. These terms, *computational* and *dynamic,* actually reflect different emphases and are not in conflict with one another. What has developed from the widespread use of computers to explore the dynamic potential of interacting systems that can process information, such as biological molecules, cells, or organisms, is a new theory of dynamical systems collectively referred to as *the sciences of complexity,* from which have developed significant branches such as artificial life.

In this book I explore the consequences of these ideas as they apply to our understanding of the emergence of biological forms in evolution, particularly the origin and nature of the morphological characteristics that distinguish different types of organism. These questions overlap those addressed by Darwin, but they focus on the large-scale, or global, aspects of biological form rather than on small-scale, local adaptations. As a result, there is no necessary conflict between the approaches, nor with the insights of modern biology into the genetic and molecular levels of organisms. These contribute to the construction of dynamical theories from which emerge higher-level properties of biological form and the integrated behavior of organisms. Conflict arises only when there is confusion about what constitutes biological reality. I take the position that organisms are as real, as fundamental, and as irreducible as the molecules out of which they are made. They are a distinct level of emergent biological order, and the one to which we most imme-diately relate.

The recognition of the fundamental nature of organisms, con-necting directly with our own natures as irreducible beings, has sig-nificant consequences regarding our attitude to the living realm. It is here that another aspect of scientific theories comes to the fore, one that is often regarded as irrelevant or secondary to the facts that science uncovers. Darwinism, like all theories, has distinct metaphorical as-sociations that are familiar from the use of descriptive terms such as *survival of the fittest, competitive interactions* between species, *selfish genes, survival strategies,* even *war games* with *hawk and dove strategies.* Such metaphors are extremely important. They give meaning to scientific theories, and they encourage particular attitudes to the processes described—in the case of Darwinism, to the nature of the evolutionary process as predominantly driven by competition, survival, and selfishness. This makes sense to us in terms of our experience of our own culture and its values. Both culture and nature then become rooted in similar ways of seeing the world, which are shaped at a deeper level than metaphor by cultural myths, from which the metaphors arise. The consequences of this perspective have emerged particularly clearly in this century, especially in the view of species as arbitrary collections of genes that have passed the survival test. The criterion of value here is purely functional: either species work or they don't. They have no intrinsic value.

I shall argue that this view of species arises from a limited and inadequate view of the nature of organisms. The sciences of complexity lead to the construction of a dynamic theory of organisms as the primary source of the emergent properties of life that have been revealed in evolution. These properties are generated during the process known as *morphogenesis,* the development of the complex form of the adult organism from simple beginnings such as an egg or a bud. During morphogenesis, emergent order is generated by distinctive types of dynamic process in which genes play a significant but limited role. Morphogenesis is the source of emergent evolutionary properties, and it is the absence of a theory of organisms that includes this basic generative process that has resulted in both the disappearance of organisms from Darwinism and the failure to account for the origin of the emergent characteristics that identify species. Many people have recognized this limitation of Darwin's vision, and my own arguments are utterly dependent on their demonstration of the path to a more balanced biology. Primary among these is the towering achievement of D'Arcy Thompson in his book *On Growth and Form* (1917), in which he single-handedly defined the problem of biological form in mathematical terms and reestablished the organism as the dynamic vehicle of biological emergence. Once this is included in an extended view of the living process, the focus shifts from inheritance and natural selection to creative emergence as the central quality of the evolutionary process. Because organisms are primary loci of this distinctive quality of life, they become again the fundamental units of life, as they were for Darwin. Inheritance and natural selection continue to play significant roles in this expanded biology, but they become parts of a more comprehensive dynamical theory of life that is focused on the dynamics of emergent processes.

The consequences of this altered perspective are considerable, particularly in relation to the status of organisms, their creative potential, and the qualities of life. Organisms cease to be mere survival machines and assume intrinsic value, having worth in and of themselves, like works of art. Such a realization arises from an altered understanding of the nature of organisms as centers of autonomous action and creativity, connected with a causal agency that cannot be described as mechanical. It is relational order among components that matters more than material composition in living processes, so that emergent qualities predominate over quantities. This consequence extends to social structure, where relationships, creativity, and values are of primary significance. As a result, values enter fundamentally into the appreciation of the nature of life, and biology takes on the properties of a science of qualities. This is not in conflict with the predominant science of quantities, but it does have a different focus and emphasis.

Darwinism sees the living process in terms that emphasize competition, inheritance, selfishness, and survival as the driving forces of evolution. These are certainly aspects of the remarkable drama that includes our own history as a species. But it is a very incomplete and limited story, both scientifically and metaphorically, based on an inadequate view of organisms; and it invites us to act in a limited way as an evolved species in relation to our environment, which includes other cultures and species. These limitations have contributed to some of the difficulties we now face, such as the crises of environmental deterioration, pollution, decreasing standards of health and quality of life, and loss of communal values. But Darwinism shortchanges our biological natures. We are every bit as cooperative as we are competitive; as altruistic as we are selfish; as creative and playful as we are destructive and repetitive. And we are biologically grounded in relationships, which operate at all the different levels of our beings, as the basis of our natures as agents of creative evolutionary emergence, a property we share with all other species. These are not romantic yearnings and utopian ideals. They arise from a rethinking of our biological natures that is emerging from the sciences of complexity and is leading toward a science of qualities, which may help in our efforts to reach a more balanced relationship with the other members of our planetary society.

acknowledgments

I am indebted to so many people for the insights that form the basis of this book that I am tempted to say that not I, but this collection of friends and colleagues, wrote it. The people who have influenced me in developing these ideas include old and new relationships, though inevitably it is the older ones that have made the more lasting impact. Among these I must mention my longest working collaboration, with Gerry Webster, which started over twenty years ago when we were both at the University of Sussex, where we discussed at length what seemed to us the deeper levels of biological meaning, involving many students in the process. Another very important influence from that time was John Maynard Smith, whose clarity of thinking about biological matters always forced Gerry and me to go further into conceptual detail than we might otherwise have done. However, John would certainly not thank me for implicating him in the direction and the conclusions of this book. The same is true of Lewis Wolpert, an old colleague whose work has been a stimulus to clarification of the alternative views on morphogenesis and evolution that are presented here. On the other hand, an equally long working friendship with Stuart Kauffman has involved a continuous convergence of ideas that, despite quite different analytical modes and emphases, has led us to virtually identical conclusions, from which I take great satisfaction and a deepened belief that we are onto a promising track.

In more recent years the influences have come from colleagues at the Open University, particularly from an ongoing dialogue with Mae-Wan Ho, whose thinking and imagination know no boundaries and continually challenge accepted limitations. I am also indebted to Steven

Rose for a dialectic about biology and society that maintained a broad perspective. We diverge more in emphasis than in objectives. For very useful comments on various chapters I am indebted to Hazel Goodwin, Jane Henry, Alastair Matheson, Jennifer Wimborne, and Françoise Wemelsfelder.

As for the many other people whose work and ideas have influenced me, I have to acknowledge them collectively here and specifically by reference in the text itself. Science is a collective enterprise, an outcome of the type of relational order that underlies all creative activity, so my own contribution is minimal. However, someone has to take responsibility for the limitations of the work, and that can only be me. My grateful thanks go to all those with whom it has been my privilege to interact over the years.

Whatever Happened to Organisms?

Something very curious and interesting has happened to biology in recent years. Organisms have disappeared as the fundamental units of life. In their place we now have genes, which have taken over all the basic properties that used to characterize living organisms. Genes multiply by making more copies of themselves; they vary by mutation; they evolve by competitive interaction, the better versions increasing in number at the expense of less useful variants. And in addition to all this, genes make organisms as a means of exploiting different environments over the face of the earth so that they can increase and prosper. Better organisms made by better genes are the survivors in the lottery of life. But behind the front that we see as the living, behaving, reproducing organism is a gang of genes that is in control. They alone persist from one generation to the next and so evolve. The organism itself is mortal, dying after a mere generation, whereas the genes are potentially immortal, the living stream of heredity that is the essence of life.

This is the biology we all know and many love, the legacy of Charles Darwin's vision of life as chance variation in the hereditary material of organisms and as persistence of the better variants via natural selection. It is a beautifully simple and elegant story of how the various types of organisms that we see about us, and the fossil forms that have left their traces, come into being and pass away, told now in terms of the adventures of their genes. This leads naturally and inevitably to the conclusion that, to understand everything that is essential about organisms, what we need to know is the information in their genes. Then we would be able to compute the adult organism in all its details of form and function, in the same way that we can predict the output of a computer from the information contained in the program it is given to read. This is why organisms have vanished from biology as the fundamental units of life, replaced by genes as their most basic and important components.

This genocentric biology is a perfectly logical consequence of the way Darwin chose to describe evolution in terms of inheritance, random variation, natural selection, and the survival of adapted species. Of course Darwin did not foresee how this story would unfold, and there were crucial changes to his ideas of inheritance that had to be made before genetics could flower into the extraordinarily powerful science that it has become in this century. There is no denying the insights that have been gained from the study of the remarkable properties of the genetic material of living organisms, DNA (deoxyribonucleic acid). But in science there is always the danger that a particular way of looking at a subject can result in tunnel vision—the assumption that it can explain everything, an inability to recognize the limitations of the approach, and reluctance to entertain other possibilities. This is what has happened in genocentric biology. It is just like the story of geocentric cosmology—the old idea that the Earth is the center of the cosmos, with all the planets and stars rotating about us. This theory worked perfectly well for a long time, and explained nearly all the astronomical observations. Then along came Copernicus, who resurrected an earlier heliocentric view the Greeks had held of the planetary system with the sun instead of the Earth at the center. Nothing of value was lost from the geocentric perspective. The detailed observations about planetary motion were simply reframed, located within a different viewpoint that was mathematically more elegant and that on subsequent testing was found to produce better predictions than the old theory.

The proposals I am going to make about biology are somewhat similar to this shift of perspective from a geocentric to a heliocentric view of the planetary system. Despite the power of genocentric biology to explain an impressive amount of biological data, there are basic areas where it fails. One of these, the most important, concerns its claims that understanding genes and their activities is enough to explain the properties of organisms. I argue that this is simply wrong. My arguments are founded on basic physics as well as on biology, and I shall use mathematics and computer modeling to illustrate the argument. But the ideas behind my position are extremely simple and straightforward. They lead to the recognition that organisms cannot be reduced to the properties of their genes, and must be understood as dynamical systems with distinctive properties that characterize the living state. This sounds pretty abstract, but it is, in fact, just as concrete as asking why the Earth goes around the sun in an elliptical orbit and getting an answer in terms of the properties of matter—gravitational attraction, the laws of motion, and so on. The position I am taking in biology could be called organocentric rather than genocentric. We shall see that organisms live in their own space, characterized by a particular type of organization. This is not a new idea, just as Copernicus's proposition was not new, but it gets clothed in a new garb of ideas that have emerged recently in physics and mathematics, as well as in biology itself.

The shift of perspective from genes to organisms does not look initially very significant. However, its consequences become more dramatic the further we follow them. It leads to a biological focus that is different from Darwin's, questioning the central importance of natural selection and adaptation as the fundamental explanatory concepts of evolution. Nothing of value in contemporary biology is lost in this shift of perspective; it simply gets reframed, reintegrated from a different viewpoint. To get started on this journey, the first thing we have to do is to look quite carefully at what genes actually do.

The Genetic Program

Genes do have some remarkable properties, but there is a definite limit to what they can tell us about organisms. Basically, genes are long strings of molecules joined end-to-end (i.e., polymers of a particular kind), with the ability to make similar long strings by a molecular copying process. The molecules of which genes are made are known as *nucleotide bases,* which come in four varieties. The genetic material of an organism is made up of sequences of these bases, arranged in a particular order and organized into the familiar double helix of DNA shown in Figure 1.1. This is the material that is considered to provide the key that unlocks all the secrets of the organism. Why? First, it can replicate itself, each strand of the double helix making a copy of the other strand because the bases match up in complementary pairs by a spatial docking process. And second, sequences of bases on one strand of the DNA can organize slightly different bases (ribonucleotides) into RNA (ribonucleic acid) polymers by a process similar to the copying mechanism. These RNA polymers (known as messenger RNA or mRNA) have the remarkable ability to organize linear sequences of amino acids into proteins, the work horses of the living cell (Figure 1.2). This is where the genetic code comes into operation, triplets of bases on the mRNA specifying particular amino acids so that a specific mRNA produces, in general, a specific protein. These are the two extraordinary jobs that the DNA can do. Of course, the DNA does not achieve these feats on its own. There is an elaborate machinery of other molecules and structures inside cells that are essential for DNA copying (replication) and for the production of mRNA

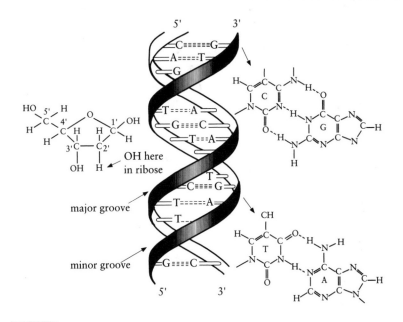

Figure 1.1   *Schematic representation of the DNA double helix and its chemical constituents.*

and protein. So the capacity of DNA to make accurate copies of itself and to produce proteins via mRNA is very much dependent upon a highly organized context: the living cell.

Within this organized context of the cell, the DNA can do more than simply produce mRNA, and, from that, produce proteins. These products can be made in a particular order in time, one type of protein being synthesized after another. This happens because the DNA itself can respond to certain types of protein (regulatory proteins) that specify which parts of the DNA (i.e., which genes) are actively producing mRNA and which are not. So the DNA can organize its own sequence of activities by producing regulatory proteins that themselves determine which proteins (including other regulatory proteins) get made next. As a result the DNA can be said to contain a program for producing a particular sequence of proteins by means of these regulatory feedback circuits. This is what we see happening during repro-

5

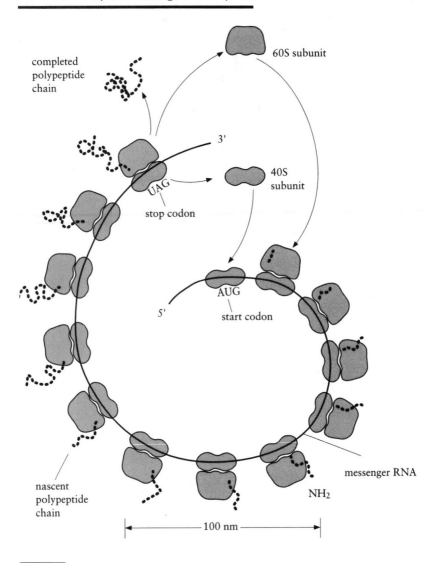

completed
polypeptide
chain

60S subunit

3'

40S
subunit

UAG

stop codon

AUG

5'

start codon

messenger RNA

NH₂

nascent
polypeptide
chain

|← — 100 nm — →|

Figure 1.2   *Synthesis of proteins on ribosomes.*

duction. Consider the simple case of an organism that is a single cell, such as the protozoan *Paramecium,* which you can find swimming about in pond water (Figure 1.3). Although it is pretty small for an organism—less than one-tenth of a millimeter in length—it is large on the scale of cells. You can see from the figure that it has a fairly complex structure, which has to be duplicated when the organism

Figure 1.3   *Structure of the protozoan* Paramecium, *known as the Lady's Slipper. The surface is covered with rows of cilia.*

reproduces and makes two cells out of one. The surface is covered with little structural elements called *unit territories,* which contain *cilia,* hairlike protrusions from the cell surface that beat like a whiplash. There are hundreds of these on each cell, all organized into regular rows called *kineties* that run from the front to the back of the organism. The coordinated beating of the cilia propels the organism with a beautiful gliding motion through the water. On one side of the cell there is a mouth, a chamber lined with cilia that direct food particles into the gullet as the organism swims about.

Inside the cell is an equally complex organization. The DNA that gets copied and passed on to a duplicated cell is organized into chromosomes that are located in the nucleus. This has its own complex structure of membranes and a basketlike network of filaments that hold the chromosomes and help to organize the patterns of activity of the DNA at different stages of the reproduction cycle. The cytoplasm is also highly organized, containing the machinery for translating mRNAs into proteins and the little chemical factories that convert the energy of simple molecules like sugar into a form that the cell uses for various types of work. There is also an elaborate and highly dynamic network of filaments that run throughout the cytoplasm; this provides

7

mechanical properties and an intricate structural organization that links all parts of the cell and is continuous with the unit territories that cover the cell surface, which are made of proteins very similar to the cytoplasmic filaments.

Paramecium reproduces by growth of the cell, duplication of all its structures, and separation of one cell into two. During this process, different genes are active at different times according to what is being duplicated—the DNA itself, the nucleus, the various components of the cytoplasm, or the structures on the cell surface. A lot of these activities go on in parallel. Others are sequential, particular proteins being synthesized in a well-defined sequence determined by the feedback circuits of gene activities changing cytoplasmic properties that in turn change gene activities. This dynamic sequence of events with its changing patterns of gene activities during reproduction is called the genetic program that directs the development of a new organism. Each species has a distinctive pattern of reproduction, which involves a particular sequence of gene activities. These sequences have now been revealed in dramatic detail, particularly in the development of more complex organisms such as the fruit fly *Drosophila,* starting from the fertilized egg and following the stages of development through to the adult. The molecular techniques that make it possible to see these changing patterns of mRNA and protein synthesis are being applied to other species such as frogs, chicks, mice, and many plant species. Each one shows distinctive features in its developmental program, but also some remarkable similarities. Excitement is running high that these sequences of molecular synthesis, the expressions of particular genetic programs, will tell us all we need to know to understand how organisms of particular kinds are produced. This is what is meant by computing an adult organism from the information in its genes. What it implies is that knowing the molecular composition of an organism and how it changes in time is sufficient to compute its form, since all that genes do is synthesize molecules. So now we have to look at what this means in terms of real physical forces to see if it actually makes sense. To do this, we have to start with a bit of physics itself.

Making Shapes

I am holding in my hand a crystal, and I tell you that it is made of carbon. Can you tell me its shape? It could be a diamond, most durable of crystals, with its beautifully regular tetrahedral form that reflects the structure of the carbon atom. But it could be graphite, whose hexagonal sheets shear off as it is rubbed over paper—a very different material, anything but durable, though great for drawing. Just possibly I've got one of those exotic crystal geodesic domes—a buckminster-fullerine, say C_{60}, with a structure reminiscent of a football with alternating hexagons and pentagons of the type predicted by that architectural wizard Buckminster Fuller. One substance, many forms. There are also crystals that have only one form, such as table salt (NaCl), in which case if you know the composition, you can specify the structure. But in general crystals are polymorphic (the same material exists in more than one crystalline form), like carbon. So we can conclude that knowing the composition of a crystal is not, in general, sufficient to determine its form. We need also to know the conditions under which the crystal was made.

What about liquids? We tend to think of these as rather formless substances compared with crystals. However, they do have characteristic forms, which are dynamic rather than static. Just think of the shape taken by water as it flows down the plug-hole of the bath—a distinctive spiral vortex, either clockwise or counterclockwise. And don't believe anyone who tells you that it flows one way in the Northern Hemisphere and the other in the Southern. Do the experiment yourself: keep track of several spiral exits from your bath and you will find that the water goes either way, depending on the movements you produce in the water as you step out of the bath. And by simply swirling the water the other way, you can reverse whichever spiral first forms. The Coriolis force that is related to the rotation of the earth is very weak, and you need special conditions to see its effects on liquid flow patterns. Only if there is no other stronger influence to initiate a vortex will the Coriolis force break the symmetry of water flow and induce a clockwise

9

spiral in the Northern Hemisphere and counterclockwise in the Southern.

Now the question arises: Could you explain the spiral vortex of a liquid flowing from a bath if I tell you its molecular composition? Suppose I tell you it is made of H_2O; or C_2H_5OH (alcohol); or Hg (mercury). Does this give you the information you need to work out the form? The composition of the liquid actually doesn't matter at all—they all form vortices under the conditions of flow down a plug-hole. So what do we need to know in order to explain these forms? We need to know the equations that describe the behavior of liquids. These depend upon the intrinsic properties of the liquid state—its properties of flow, incompressibility, viscosity, adhesivity, and so on; not its composition. These properties of fluids are described by particular equations that were derived independently in the nineteenth century by the mathematicians Claude Louis Marie Henri Navier, a Frenchman, and by George Gabriel Stokes, an Englishman, who are together immortalized in the Navier-Stokes equations. When these are solved for the particular conditions of water flow from a bath—a container with a hole, filled with fluid acted upon by gravity—out pop clockwise and counterclockwise spirals as the stable solutions of motion. To determine which of these occurs, we need to take account of the specific conditions acting on the liquid, such as the initial motion of the fluid due to a person stepping out of the bath or swirling the liquid deliberately in one direction, or the action of the Coriolis force. And to specify the pitch of the spiral and its rate of flow, we need to know such factors as the temperature, the strength of the gravitational field, and also the specific gravity and viscosity of the liquid, which is where composition finally comes in as an influence—pretty far down the list, though nevertheless a contributing feature.

All right, you are no doubt thinking, enough of this physics and mathematics. What has this got to do with the shapes of organisms, which are neither crystals nor liquids? But organisms do obey the laws of physics, and in fact crystallization is a very important contributor to the process of making shapes in organisms and their parts. I am not thinking only of bones and teeth, which are crystalline materials. I am also considering structures such as the lens of the eye, which is a crystal made out of protein! Despite their large size (thousands of times larger than a carbon atom, which is just one of their components), proteins make beautiful crystals. Given their complexity, we might expect to find polymorphisms in crystal proteins as well as in crystals of carbon. And here is one example.

Bacteria swim about in water in much the same way as *Paramecium,* using the whiplike action of filaments that extend out of the cell. These filaments are called *flagella* (singular, *flagellum*), and there are usually two per cell rather than the hundreds of cilia on *Paramecium,* which is also much larger than a bacterium (about ten thousand times larger by volume). The bacterial flagellum is made of an elongated crystal of a single type of protein called *flagellin,* consisting of many molecules that fit together in an orderly array. This crystal isn't rigid, because the individual protein molecules are flexible, and they can also move slightly relative to one another without losing their structure. It was discovered that in a particular type of bacterium called *Salmonella* (one of the nasties involved in food poisoning) there are two different shapes of flagella. One is the common form, which is called *wavy* because the flagellum has a natural undulating structure. A rarer mutant form has a shorter wavelength and is called *curly.* (See Figure 1.4). Suppose we collect a whole lot of wavy flagella and separate them into their constituent protein molecules, which can be done without damaging the protein, flagellin. Now we let it recrystallize to form flagella-like structures. The result is wavy flagella, as you would expect. If we do the same with curly flagella, we again get the curly form when the protein recrystallizes into filaments. Now we can do further experiments to test for polymorphism. The flagella can be broken up into fragments rather than being disaggregated into single flagellin molecules. Suppose we take some fragments of curly flagella and add them to a solution of pure, molecular flagellin from wavy flagella. The result? Curly flagella. So wavy flagellin can make either type of flagellum! On its own, it spontaneously makes wavy flagella.

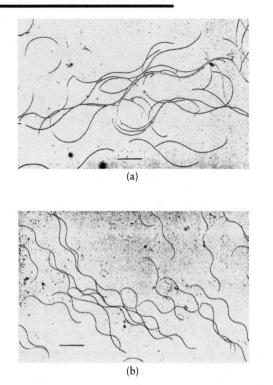

(a)

(b)

Figure 1.4   *Structure of (a) wavy flagella, and (b) curly flagella.*

But if curly fragments are present, acting as crystal "seeds" that direct the crystallization of wavy flagellin, then the curly form can be produced. So wavy flagellin is polymorphic, like carbon. Therefore it isn't enough to know the protein composition to determine the shape of the crystalline form, the flagellum. We also need to know how the crystallization process was started—that is, what seeds were present.

What about the reverse experiment: curly flagellin (protein molecules) and wavy fragments (seeds)? The result is curly flagella. In this case, the protein determines the form, not the seed. And in fact we might have predicted this if we thought about what it means to get a mutant form of flagella. When a bacterium is getting ready to divide into two cells, it has to make new flagella. It does this by copying the flagella that are already there, that is, it uses the existing flagella as seeds for the new ones, made out of newly made flagellin modules, so that the new flagella are made in the right place in the duplicating cells. Suppose the gene that codes for flagellin changes (mutates) so that a curly type of flagellin is produced instead of the normal wavy type. If this curly flagellin made a wavy flagellum under the influence of the normal wavy flagellum acting as a seed, then the mutation would never show up as a curly flagellum. In order for a new form to be produced, the mutant protein has to override the seeding influence of the wavy flagellum already present in the cell. So we could have predicted the result of wavy seeds plus curly flagellin, producing curly flagella. This example gives us a further warning: We have to be careful about the molecular interpretation of a mutation. Once the importance of crystal seeds has been recognized, it suddenly becomes evident that cells can have mutations that do not start with a change in a gene. Something can happen to the organization of the seed or its equivalent that is independent of genes. To give an example of this, let's consider some fascinating observations that were made with our little friend *Paramecium*. American biologist Tracy Sonneborn noticed one day that among his cultures of normal *Paramecia* was a cell with a form he called a "melon stripe" because one of the rows of cilia was reversed, resulting in a stripe on the cell that reminded him of the pigmentation stripes on a melon. He isolated this cell and let it grow and divide to produce others. All its progeny had the same reversed row of cilia. So he had a mutation—it bred true. What caused it?

Sonneborn was a highly skilled experimentalist. Despite the very small size of *Paramecia,* he could operate surgically on them, for example, removing patches of cilia from the surface. So he cut out a row of cilia from a normal *Paramecium* and put it back in reversed orientation. This healed, and he had created a melon-stripe cell. What would happen when it divided? He had good evidence that its genes were unchanged by the operation, so genetically it was a normal *Paramecium*. Only its body form had been altered. If genes determine form, then new cells should be normal, and the effects of the operation should be eliminated. But that is not what happened. All the progeny of the altered cell had the same melon stripe, the same reversed row of cilia. So here is a case of a mutation, produced by an operation, that is inherited via body structure, not via the genes. It is a case of cytoplasmic rather than nuclear inheritance. And according to the principles of crystallization and the effects of crystal seeds, this is not in the least surprising. The reason is that when *Paramecium* is making new unit territories and cilia in preparation for dividing into two cells, it uses the ones already there as "seeds" to direct the assembly of the proteins that make up these structures, just as bacteria use their flagella as crystallization seeds to assemble flagellin into new flagella. The reversed row of cilia therefore gets copied when the melon-stripe cell develops into two cells. So this is an example in which a mutation arises from change in a cell structure rather than in a gene. The same molecules are being produced by normal genes, but they are assembled into unit territories and cilia that simply have a reversed orientation in one row. Of course, there are plenty of genetic mutants in *Paramecium* as well, and they are much more common than the cytoplasmic variety (which can arise spontaneously by errors during division of the cell rather than by deliberate operation). But it is clear that inheritance does not depend just upon genes; it also depends upon any cytoplasmic organization that is transmitted from one generation to the next and can exist in different stable forms. Organisms can have both genetic *and* cytoplasmic inheritance.

The type of reproduction that *Paramecium* undergoes is called *vegetative,* or *asexual,* reproduction: one organism gives rise directly to another, with a continuity of body structure from one generation to the next. This type of reproduction is not confined to single cells. There are plenty of multicellular organisms, plants as well as animals, that normally reproduce this way. A lot of plant species have underground or overground propagators that generate new plants, such as the roots of *Convolvulus* or strawberry runners. Many animal species normally reproduce in a similar way to *Paramecium,* with a new organism being produced directly from the parent. Sonneborn used one such species, a worm called *Stenostomum,* to show that it had the same type of cytoplasmic inheritance as he had demonstrated in *Paramecium*. And there are many other animal species with the same properties.

Given these examples, and many others, that show how important the actual body of the parent organism is in reproduction, why is there such an emphasis on genes? There is a perfectly good reason. In sexual reproduction, two cells only—the egg from the female and the sperm from the male—come together to produce the cell from which the new organism is made. The only thing that the male contributes to this union is a set of chromosomes, carrying the genes that influence how the new individual will develop and the shape of its body. A single mutant gene can result in six toes on a kitten instead of the normal five, or two pairs of wings on a fruit fly instead of the usual single pair (Figure 1.5), or a snail with a right-handed instead of a normal left-handed spiral shell. So a change in one gene can make a big difference to the shape of an organism, or indeed to any other inherited property. This is a very important observation, and a lot has been made of it. But the conclusion is often drawn that the genes themselves, through their products, contain the key to understanding how all the detailed properties and structures of organisms are made, so that all we need to know is what the genes are doing in order to explain how organisms get their shapes. This is a primary motivation for an international endeavor, the Human Genome Project, to get a complete

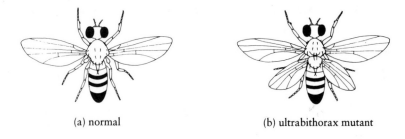

(a) normal   (b) ultrabithorax mutant

Figure 1.5   *Comparison of (a) normal fruit fly* Drosophila *with (b) an ultrabithorax mutant, which has two pairs of wings instead of the normal pair.*

catalogue of all the genes in a human being. As expressed by one of the participants in this project, C. Delisi:

> This collection of chromosomes in the fertilized egg constitutes the complete set of instructions for development, determining the timing and details of the formation of the heart, the central nervous system, the immune system, and every other organ and tissue required for life.

The logic that leads to this very strong statement runs basically as follows. Because we know that a change in a single gene is enough to cause a change in the structure of an organism, genes must contain all the information for making that structure. If we can get that information, we'll understand how the structure is made. Sounds plausible. But does it hold water? Let's go back to the bath to find out. Suppose the water is flowing down the drain with a counterclockwise vortex. You then swirl the water about with your hand and produce a clockwise vortex. The motion of your hand caused the difference. Does it therefore explain the spiral flow? Certainly not. We need to go back to Navier and Stokes to get the explanation of why liquids flow down drains with unidirectional spiral motion. But something has to determine which of the possible directions is selected—the motion of the water you produce as you step out of the bath, or the Coriolis force, or the movement caused by your hand. This is like the seed that can specify the structure of a crystal, but in this case it is a liquid seed, what is called an *initial condition* for the equations, selecting one of the possible solutions. I am going to present evidence that this is what genes are doing in making a difference to the structure of an organism. They can select or stabilize one of the alternative forms available to organisms. Of course, there are many more of these alternatives for organisms than there are for liquids or crystals. I go into this in more detail in chapters 4 and 5, where I discuss the type of dynamical system that an organism is and how it can spontaneously generate the complex and beautiful structures that we see in plants and animals. But before embarking on that fascinating exploration, I have to tell you more about how biology came to be as it is, so that we can more clearly understand how it can change. In the next chapter, I take a step back into history to get some perspective on the origins of current ideas so that we can see what basic assumptions need to be altered before biology can set out on a new path. This does not deny anything of value in contemporary biology; rather, it builds on it but shifts the focus in a quite intriguing way.

How the Leopard Got Its Spots

Darwinism has been phenomenally successful as a scientific theory. It is often ranked with Newton's theory of motion as one of those rare, enduring achievements of the scientific imagination. Darwin's vision of evolution by random variation of inherited characteristics in organisms and by selection of the fitter variants is so simple and so convincing that once you have grasped it, you feel you are in possession of a universal truth. And indeed, this idea, or simple variations of it, tend to be applied to everything in our culture that is complex and changing—to the evolution of social and economic systems, to competition and survival in the business world, even to the development of new ideas themselves. When that kind of phenomenon occurs, you can be sure that there is more involved than just a good idea. That good idea is also resonating deeply with one of our cultural myths. By myth I do not mean anything in the least belittling or lacking in reality value. On the contrary, myths are more real, more relevant, and often more permanent than what are taken to be facts. Facts are considerably less durable in science than you might expect, as the following example illustrates.

In Darwin's time, physicists had made a calculation about the age of the earth on the basis of its temperature and the rate of cooling of solid bodies. The result was the "fact" that the earth could not be older than a few million years—longer than the Bible said, but much shorter than Darwin needed for evolution to have occurred by small cumulative variations and natural selection. But he stuck to his theory, and later the fact about the age of the earth changed. Radioactivity was discovered, and it became evident that radioactive decay is a significant source of heat within our planet, so that the rate of cooling is much slower than was originally assumed. The estimated age of the earth shot up to several billion years, plenty of time for evolution, though Darwin was dead by the time this factual correction occurred. His theory of evolution is outlasting a lot of facts, which, after all, are just theoretical interpretations of available evidence. But the cultural myth with which Darwin's theory resonates is actually much older than science itself, and it is still much alive, as we shall see.

Scientific theories are based on decisions about the basic problems that characterize some area of inquiry. Before Darwin, biology was largely about morphology—the structure of organisms. It was observed that different types of organisms could be related to one another based on similarities and differences in their structures, particularly their bones and other hard parts such as shells, which survive as fossils, making it possible to compare living and extinct species. History was therefore an important dimension in biology, but the systematic comparison of organisms and the construction of a classification scheme of relationships among species was largely a logical enterprise, like classifying the elements in physics into the periodic table. The difference in biology is that species come and go, arise and become extinct, as we know only too well from current rates of extinction, whereas most of the elements are quite permanent, stable inhabitants of our world. So biology involved history, but what was important was the structural relationship among species, not their historical origins. Species were divinely created, after all, as the Church insisted and as most people believed, and in the early years of the nineteenth century, evolution was an idea that appealed primarily to a few radical French thinkers.

Darwin belonged to an age that had discovered historical explanations and was becoming preoccupied with change and the reasons for it, as Europe experienced increasing rates of social and political transformation. In biology, the continuous accumulation of fossils made their history more and more problematical: Why did all those species go extinct? Why do the ones we see about us survive? Darwin cast about for a solution. He had a vast acquaintance with the behavior of living species in their natural habitats and was thoroughly grounded in the comparative study of species morphology. What was lacking was a dynamic perspective that could make sense of all these "facts" in the context of history. Darwin took two concepts that were already part of biology and added a third crucial ingredient that made his evolutionary theory rise in the oven of creative insight and make a very palatable product to many of his contemporaries, though it stuck in the craw of many others. The two concepts he found at hand were the notion of adaptation (organisms are suited to their habitats) and inheritance (offspring resemble their parents). One of the most prominent and influential books on adaptation in Darwin's day came from what might appear to be an unexpected source: the Church. A theologian, William Paley, had written the comprehensive volume *Natural Theology; or, Evidences of the Existence and Attributes of the Deity, collected from the Appearances of Nature.* This described in detail the remarkable and diverse ways in which organisms are adapted to their environments, which was interpreted as a major proof of God's existence, since only an intelligent creator could have generated such goodness of fit between the organism and its habitat. How else could one account for the beautifully designed relationships between insects and flowers, one getting food and the other achieving cross-pollination, to the mutual benefit of each? Or the exquisite aerodynamic precision of bird wings that results in the soaring flight of the hawk? And the miracle of the eye that allows it to spot the movement of a vole at five hundred feet? Paley provided copious material on this theme, over a great range of species, and Darwin accepted the phenomenon of adaptation as one of the primary facts of biology. He had studied theology at Cambridge in preparation for taking Anglican orders, so this worldview was utterly familiar to him, deeply embedded in his psyche. However, invoking a transcendental God to create these forms was not an acceptable scientific solution to how they came into existence, as Darwin was fully aware from lively conversations with free thinkers and from reading the works of iconoclastic French biologists such as Étienne Geoffroy Saint-Hilaire and Jean Baptiste de Lamarck. Darwin pondered alternatives.

Inheritance was the other concept that Darwin made use of in developing his ideas about evolution. Why do offspring resemble their parents? The dominant theory of Darwin's time was that organisms have the capacity to adapt to their environments, and so parents pass their adaptations on to their children. For example, animals produce more hair in cold weather. The woolly mammoth was believed to be a result of adaptive changes in this property: as the climate gets colder during a glacial age, parents grow more hair and transmit extra hereditary potential for hair growth to their offspring, so they can produce even more hair. This is known as Lamarckian inheritance after Lamarck, who made the idea prominent in the early years of the last century. It was a view that Darwin came to accept. So this could explain adaptation. But Darwin needed an idea to account for extinction. And he found it in the notion of competition for limited resources. In any environment there is only so much food to go around, only a limited number of safe places to make a home: caves, hollow logs, holes in trees, burrowing sites. But why don't individuals recognize these limitations, becoming adapted by producing the appropriate numbers for the environment? Darwin assumed that at this point individual adaptation fails, but *population* adaptation succeeds. Organisms have the potential for exponential growth, offspring being more numerous

How the Leopard Changed Its Spots

than their parents so that the population is constantly tending to increase from generation to generation. Darwin got this idea from a study of population growth by his contemporary Thomas Malthus. Since resources are always limited, but populations tend always to increase, there is inevitably a conflict for what is available, and the less well-adapted members of the population will die. There is always a lot of random variability in populations—differences in size, color, strength, speed, rate of growth, and so on. Among these, the surviving organisms are, inevitably, the better adapted to their circumstances.

So here was the third ingredient that gave Darwin the recipe for a dynamic theory of evolution that accounted for both adaptive change and extinction. As the environment changes (and of course this includes other organisms), the pressures of survival force changes on populations. This is natural selection, the emergence of new, adapted types of organisms and the extinction of those that fail to change adequately. The different types of organisms are just arbitrary groupings of continually changing populations into convenient categories such as plants and animals, animals with and without backbones, animals with and without feathers, animals with and without a placenta for bearing young internally, and so on. These categories are a result of the history of adaptive response to changing environments and the accidents of heredity that confer better survival capacities on some rather than others. Now history begins to play a really significant role in evolution. If we want to understand how the vast range of biological forms has come into existence, we have to track the historical relationships of the different types of organisms back through time to see how they evolved from one another by adaptive radiation, and to reconstruct the environmental changes that occurred over geological periods—the ice ages; the weathering of mountains and changes in the salt content of the oceans; the movements of the continental plates; the formation and disappearance of oceans, seas, and lakes. So we enter a new world, one of perpetual change in which history, heredity, and adaptation through competitive interaction are the ingredients of an evolutionary biology. This was Darwin's vision:

a magnificent, inspiring unification of the biological realm in which all organisms are joined together in the tree of life on earth, rooted in one common origin, the branches and twigs of which are the diverse forms that have evolved by adaptation to different circumstances.

Two aspects of this vision provoked intense reaction from the nineteenth-century establishment as represented by the Church. One was the abolition of the creative power of the Deity in designing the forms of life: living matter itself was given this creative potential. The other was the union of all organisms—including humans—within a single evolutionary process. With one stroke, God and humans were toppled from privileged positions in the scheme of things. That is asking for trouble, and Darwin got it. However, he had some extraordinarily able allies who loved the arena of public debate and thrived on dispute, unlike Darwin, who preferred to keep a low profile and work away at the evidence for and against his theory. Foremost among his supporters was the eminent biologist Thomas Henry Huxley, who dominated many crucial encounters through a combination of devastating debating skill, complete familiarity with the subject, and no hesitation to exploit personal details about his opponents that were irrelevant to the issues being debated. For instance, one of Darwin's most cogent critics was St. George Jackson Mivart, who marshalled an impressive array of evidence that adaptation to the environment by natural selection is a quite inadequate basis for explaining species morphologies. The marsupial mammals of Australia, for instance, have evolved independently of their European relatives, the placental mammals, and have experienced a different range of environments, yet they have evolved into strikingly similar species in the forms of wolves, cats, jumping and ordinary mice, flying and nonflying squirrels, and even a placental kangaroo rat that has the same tooth pattern as the kangaroo. Mivart produced a host of such examples. However, it so happened that he had undergone conversion from the Anglican to the Roman Catholic Church. Although this was hardly an issue of significance in relation to whether or not organisms are adapted to their habitats, Huxley argued that Mivart's objections to Darwinism arose from his opposition to evolution as a matter of theological principle, thus disqualifying his views on adaptation. There is more to scientific debate than facts and theories. And we shall see that, despite fierce encounters with theologians, Darwinism continued to function within a cultural myth that it shares with all Christian theologies.

The Death of the Organism

You may have noticed an inconsistency in Darwin's theory of evolution. He believed that parents could pass on to their offspring adapted characteristics that they had acquired during their lifetimes. This is Lamarckian inheritance. If all organisms in a population have this capacity, then they are all equally able to evolve, and there is no need for any selection of fitter variants—all will be equally fit. There will still be competition if more organisms are produced than can survive, but survival is a pure lottery among equals. The person who noticed this inconsistency and worried about it a hundred or so years ago was German zoologist August Weismann. He realized that the crucial question concerned inheritance. Was Lamarck right? Like Darwin, Weismann originally believed so. Then he did a number of experiments and investigated the process of inheritance in insects. He became convinced that Lamarck had got it wrong. The concept of inheritance that he put in its place laid the foundations for the spectacular revelations about DNA that have emerged in our time. Oddly, Weismann also got it wrong, as we shall see. But he set biology off on an extremely fruitful track. In science, after all, it is never a case of being permanently right, because sooner or later all facts and theories change (including "laws"). That is the dialectical process of understanding. Weismann's idea was based on correct, though limited, observation; the significance of his contribution is that it was appropriate to the development of a very valuable insight in biology. What is more, it resonated with the same cultural myth as Darwin's theory.

Weismann was interested in how insects develop from egg to adult form. He discovered that there is a particular part of the egg that forms the reproductive cells (eggs and sperm) and so is specifically involved in inheritance. Figure 2.1 shows the earliest stages of this process in the fruit fly *Drosophila*. The fertilized egg has a single nucleus that divides into two; these then divide to give four nuclei, and so on until there are 128 nuclei. These all migrate to a position just under the cell

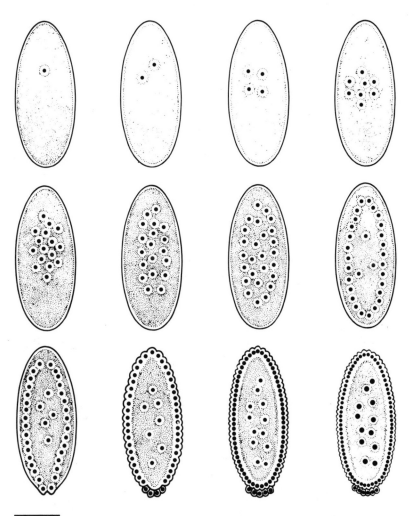

Figure 2.1   *The earliest stages of* Drosophila *development: the synchronous divisions of the nuclei in the egg; migration of the nuclei to the cell surface; and formation of the germ cells at the posterior pole of the egg. The germ plasm is the cytoplasm where the pole cells form.*

membrane, forming a layer of nuclei that go on dividing. Then, at the posterior pole of the developing embryo, the first embryonic cells form: cell membranes grow around the nuclei and separate them from the rest of the embryo. These are known as the *pole cells,* and it is they that will become the *germ cells,* the eggs and the sperm in the reproductive organs, producing the next generation. Weismann called this part of the embryo the *germ plasm.* The other part of the embryo, which produces the rest of the body (the soma) of the adult organism, he called the somatoplasm. This body dies at the end of the organism's lifespan, whereas the germ plasm gets transmitted to the next generation, and from that one to the next and so on, so that it is potentially immortal. Weismann knew from the work of others that the most likely part of the germ plasm to be involved in inheritance is the chromosomes within the nuclei. He identified these as the physical carriers of the hereditary determinants that are passed on from generation to generation. He did not know about genes, because Gregor Johann Mendel's experiments, though carried out some thirty years or so before Weismann's own work in the 1880s and 1890s, were not rediscovered until 1900.

The distinction between a mortal somatoplasm and an immortal germ plasm is shown schematically in Figure 2.2. The crucial idea about inheritance is that the hereditary determinants that are carried by the germ cells from generation to generation contain the instructions

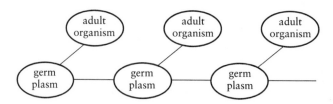

Figure 2.2   *Weismann's description of the properties of the germ plasm, which can reproduce itself indefinitely through future generations and which directs the formation of the adult organism from the somatoplasm. The adult is mortal and cannot influence the germ plasm.*

for making the adult organism. The role of the adult is to provide a vehicle for the transmission of these instructions to the next generation by making use of a particular habitat for growth and reproduction. Any variations that occur in the hereditary instructions will result in corresponding variations in the properties of the adult organism so that evolutionary change occurs via modifications in the hereditary material. It was here that Weismann introduced the hypothesis that resolved Darwin's inconsistency. He proposed that, whereas the hereditary information in the germ cells directs the formation of the adult organism (the soma) through the process of development, the soma cannot influence the hereditary information in the germ cells. He introduced a block, *Weismann's barrier,* that prevents information from flowing back to the hereditary material in the germ cells from the soma. Organisms may produce more hair in response to colder weather, more muscle as a result of lifestyle, more pigmentation in response to sunshine, or more hemoglobin in response to altitude. But these acquired characteristics are not transmitted to the next generation via information in the germ cells. This makes Lamarckian inheritance through the germ cells impossible. However, Lamarckian transmission is possible via the soma in cases of reproduction that involve continuity of the body of the organism from one generation to the next. We have already seen examples of this in chapter 1 in the case of organisms such as the unicellular *Paramecium* and the worm *Stenostomum,* and there are plenty of others. But for any species reproducing sexually, where the only cells involved in reproduction are the germ cells, Weismann's barrier forbids the inheritance of acquired characteristics. Darwinism thus acquired a theory of inheritance that was consistent with the principles of natural selection, and this was fiercely defended. Any whiff of Lamarckist sin was quickly and vehemently dealt with. It came to be the great heresy, and it still is.

Weismann saved Darwinism from inconsistency by separating organisms into two distinctly different parts, one mortal and transient (the body) and the other potentially immortal, the transmitter of hereditary instructions (the chromosomes in the germ cells). The theory provided the perfect context for the science of genetics that developed very rapidly after the rediscovery of Mendel's work on inheritance in 1900, with the demonstration that the hereditary factors obey well-defined rules of transmission from generation to generation. A great advantage of Weismann's scheme was that genetics could develop without understanding how genes act during development. All that was necessary was to see the effect of genes on the characteristics of the adult organism, so that their presence or absence could be inferred— tall plants signified the presence of particular genes affecting height, red flowers signified the presence of particular genes making red pigment, and so on. *How* genes acted to produce their effects could be put aside while the rules of their segregation and transmission were worked out. The main stream of life flowed with the potentially immortal germ plasm, carrying the genes whose spontaneous variations by segregation, combination, mutation, and other sources of change produced the character variations in organisms on which natural selection acted, resulting in evolutionary change. Organisms became nothing but the vehicles for genes. Genetics went from strength to strength, and in the 1950s the nature of the hereditary material was identified: the double helix of DNA (deoxyribonucleic acid) was deduced by James Watson and Francis Crick, based on the pioneering work of the biochemist Oswald Theodore Avery and on the insights of Rosalind Franklin and Maurice Wilkins into its basic structural properties. From this came the revelation of how DNA achieves its dual functions of self-replicating and making messages for protein synthesis. These two activities explained in precise molecular terms the two aspects of Weismann's theory (Figure 2.2). Continuity of the hereditary material of the germ cells from generation to generation was achieved by the self-copying capacity of DNA; the instructions for making an organism via the process of development came from DNA through its capacity to make messenger RNA, and hence protein, the stuff of which the soma, the body, is made. So we can identify the DNA as the potentially immortal hereditary stuff of life, while protein is the mortal material that makes up the transient body of the organism, as shown in Figure 2.3. Weismann had anticipated the whole unfolding of molecular biology in this century. What a stunning bit of prescience!

The Myth behind the Metaphors

The story of the success of genetics and molecular biology in making sense of evolution at the basic level of molecules and genes has been told many times, but one of the most interesting and influential versions is contained in Richard Dawkins's books *The Selfish Gene, The Extended Phenotype,* and *The Blind Watchmaker.* These are powerful, persuasive presentations of genocentric biology, whose clarity of exposition is enhanced by the metaphors used to vivify the science. These metaphors are themselves revealing, because they are not just arbitrary coloring applied to a monochromatic scientific discourse. They actually belong in a very direct and intimate way to the ideas, informing them with relevant associations and reflecting the deeper myths with which the science resonates.

First, there is the selfish gene metaphor. Of course, genes are not selfish. But they behave as if they were. The capacity of DNA to make copies of itself gives it the property of a "replicator" that is seen as the basis of organisms' capacity to increase exponentially and transmit

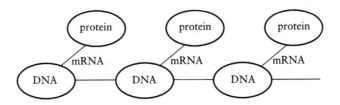

Figure 2.3   *The relationship between DNA and protein is like that between germ plasm and somatoplasm: DNA replicates itself and so can be perpetuated in future generations, whereas protein, produced under the direction of DNA (via mRNA), has a limited lifetime and cannot change the information coded in the DNA.*

genes from generation to generation. "Better" genes make "better" characteristics, that is, the organisms with these characteristics have a better chance of surviving and transmitting these genes to their progeny. So genes, via the organisms they produce, are constantly "trying" to leave more copies of themselves. In this process organisms compete with one another for scarce resources, as Darwin described. But in *The Selfish Gene* Dawkins shows how not only competitive but also apparently cooperative behavior of organisms, such as parental care and social organization, are really the result of every gene looking out for itself and trying to increase its numbers. Then, at the end of the book, comes an interesting surprise: "We are built as gene machines and cultured as meme machines, but we have the power to turn against our creators. We, alone on earth, can rebel against the tyranny of the selfish replicators." So human beings, alone among the species, have the possibility of "deliberately cultivating and nurturing pure, disinterested altruism." Culture can triumph over nature in this one species. (It is not entirely clear how an organism, constructed by selfish genes for their own perpetuation by whatever means they can devise, can escape their cunning influence. How can such autonomy, such freedom to choose altruistic behavior, arise unless it, too, is of benefit to the genes that generate this property, in which case again altruism is really self-interest? However, I am not going to pursue these questions here; they will come back in another context in chapters 6 and 7.)

Dawkins's description of the Darwinian principles of evolution can be summarized as follows:

1. Organisms are constructed by groups of genes whose goal is to leave more copies of themselves. The hereditary material is "selfish."

2. The inherently selfish qualities of the hereditary material are reflected in the competitive interactions between organisms that result in survival of fitter variants, generated by the more successful genes.

3. Organisms are constantly trying to get better (fitter). In

a mathematical/geometrical metaphor, they are always trying to climb up local peaks in a fitness landscape to do better than their competitors. However, this landscape keeps changing as evolution proceeds, so the struggle is endless.

4. Paradoxically, humans can develop altruistic qualities that contradict their inherently selfish nature by means of educational and other cultural efforts.

Does this list look familiar? Here is a very similar list of principles from another domain:

1. Humanity is born in sin; we have a base inheritance.
2. Humanity is therefore condemned to a life of conflict and
3. Perpetual toil.
4. By faith and moral effort humanity can be saved from its fallen, selfish state.

So we see that the Darwinism described by Dawkins, whose exposition has been very widely (but by no means universally) acclaimed by biologists, has its metaphorical roots in one of our deepest cultural myths, the story of the fall and redemption of humanity. Dawkins did not invent this evolutionary story; he just tells it with great care and inspiration, in terms that clarify the underlying ideas of Darwinism. And what we see so clearly revealed is a myth with which we are all utterly familiar. So now we can begin to see what Darwin did, from a different perspective. He certainly was guilty of heresy regarding the theological version of the fall/redemption story, because he had no place for God or the human spirit in his version of the origin of species. For him, creativity is in living matter itself. The capacity of life to generate an indefinite diversity of forms, of species, is intrinsic to this state of organization of matter, and no transcendental spirit is required to animate it and give it form. But once this crucial step had been taken, the rest of the story remained much the same as before: competition, struggle, work, and progress. Darwin saw this as the path to civilization and to culture, as revealed in an interesting comment he made about the natives of Tierra del Fuego, whom he encountered during his South American voyage on the *Beagle*. He noticed that the possessions the Fuegians had acquired from members of the ship's company had been divided equally among all. Darwin saw this as cooperative savagery and remarked, "The perfect equality of all the inhabitants will for many years prevent their civilization. Until some chief arises who by his power can heap up possessions for himself, there must be an end to all hopes of bettering their condition." Of course there is nothing specifically theological about such views, except for the intimacy of the relationship between the work ethic in our culture and the notion of redemption by good works. Darwin's theology was more of the natural kind, the view espoused by Paley that adaptation of species to their habitats is one of the most striking features of the biological realm. But instead of seeing this as evidence of the Creator, Darwin viewed this as evidence of the power of natural selection, which became his creative principle. Darwin's heresy was to materialize these theological propositions, but otherwise the basic set of ideas and metaphors remained intact. These can be seen in any popular exposition of Darwinism, and their emergence in Dawkins's writings is simply a recent example. (An excellent examination of the way modern biologists use myths in their writing is Howard Kaye's book *The Social Meaning of Modern Biology*.)

How did Weismann's contribution fit into these metaphors? We have seen that he divided organisms into a mortal body and a potentially immortal germ line, its hereditary essence, that is transmitted from generation to generation. Again, this dualism is very familiar. It parallels the mortal body and the immortal soul of many religions. And we all know which part is more important. The hereditary material is clearly the essence of the organism, and genocentric biology is firmly established on metaphorical and theological foundations. This essence becomes Dawkins's replicator, which is the primordial self-replicating organism. The body of the organism, which to the naive observer seems to be the main part, is really just packaging for the hereditary essence. Weismann did not himself go quite this far, but he made the distinction perfectly clear when he spoke of the germ plasm as "living in a distinct sphere from the somatoplasm."

Now the point of this exercise is not to conclude that there is something wrong with Darwin's theory because it is clearly linked to some very powerful cultural myths and metaphors. All theories have metaphorical dimensions, which I regard not only as inevitable but also as extremely important. For it is these dimensions that give depth and meaning to scientific ideas, that add to their persuasiveness and color the way we see reality. Science, after all, is not a culture-free activity. The point of recognizing this and the influences that act within current Darwinian theory is simply to help us stand back, take stock, and contemplate alternative ways of describing biological reality. The choices that Darwin and Weismann made in constructing a conceptual scheme for understanding evolution were neither arbitrary nor inevitable. They came from previous ideas about life and creation, but they were transformed in particular ways. There are biologists who take the view that Darwin's theory of evolution is so rock solid, so well formulated and complete in its essentials, that no alternative can be contemplated. The maintenance job on this theory can be handed over to historians and philosophers, now that biologists have done the spade work and established its permanent foundations. Such confidence is always interesting, for it reflects the power and persuasiveness of a particular "way of seeing" that has cultural roots as deep as Darwinism.

However, no scientific theory is permanent, and the rest of this book tells a rather different story, another transformation of current ideas. The reason for making this change is not merely an unease with the metaphorical structure of Darwinism; it is with the science. Some of the basic assumptions that underlie the conceptual structure of the present view of biology are inconsistent with the evidence. Inconsistency in science is no great sin, as we have seen—it is a spur to clarification. But I see a series of inconsistencies adding up to a need for major revision.

As the Spots Disappear, So Does the Leopard

Nothing is inevitable in science; it is informed by a whole set of interlocking levels of meaning and understanding that are selected to make "best" sense of the world according to the viewer. This does not mean that anything goes, that any set of ideas is satisfactory as a way of seeing. Rather, it is like the process of making sense of one of those ambiguous figures that can be seen in either of two ways. Both are "real," and we can switch back and forth once we recognize them. But for particular purposes, one is better than the other. It is up to us to use scientific ideas for particular purposes, and to be clear about why we choose one rather than another. So I am not going to argue for a new and better paradigm in biology that is going to out-compete the previous one. That is the authoritarian mode of science that characterized modernity. We are now into a postmodern age where things are beginning to happen differently, and none too soon. This is where you exercise your judgment. You can decide to stick with the current view, which has its advantages, or to contemplate other paths, one of which I now describe.

I will start with a few of the inconsistencies that I see in the biology that has emerged from the ideas of Darwin and Weismann, together with the work of Mendel and a whole army of contributors in this century. Some of these inconsistencies were already considered in chapter 1, and more will be considered later.

    1. The proposition that "the collection of chromosomes in the fertilized egg constitutes the complete set of instructions for determining the timing and details of the formation of the heart, the central nervous system, the immune system, and every other organ and tissue re-

quired for life" (C. Delisi, 1988) is incorrect. These instructions, which define a genetic program, can determine the molecular composition of a developing organism at any moment in its development, but they are insufficient to explain the processes that lead to a heart, a nervous system, a limb, or any other organ of the body. The reason is, as we saw in chapter 1, that knowing the molecular composition of something is not, in general, sufficient to determine its form. This follows from basic physics. We also need to know the principles of organization that are involved in the system to explain what forms it can take. Then we can understand how factors such as molecular composition influence the development of a particular form. So the morphology of organisms cannot be explained by the action of their genes. One of the leopard's distinctive spots fades away.

2. The DNA of an organism is not self-replicating; it is not an independent "replicator." The only way in which the DNA can be accurately and completely replicated is within the context of a dividing cell; that is to say, it is the cell that reproduces. In a classic experiment, Spiegelman in 1967 showed what happens to a molecular replicating system in a test tube, without any cellular organization around it. The replicating molecules (the nucleic acid templates) require an energy source, building blocks (i.e., nucleotide bases; see Figure 1.1), and an enzyme to help the polymerization process that is involved in self-copying of the templates. Then away it goes, making more copies of the specific nucleotide sequences that define the initial templates. But the interesting result was that these initial templates did not stay the same; they were not accurately copied. They got shorter and shorter until they reached the minimal size compatible with the sequence retaining self-copying

properties. And as they got shorter, the copying process went faster. So what happened was natural selection in a test tube: the shorter templates that copied themselves faster become more numerous, while the larger ones were gradually eliminated. This looks like Darwinian evolution in a test tube. But the interesting result was that this evolution went one way: toward greater simplicity. Actual evolution tends to go toward greater complexity, species becoming more elaborate in their structure and behavior, though the process can also go in reverse, toward simplicity. But DNA on its own can go nowhere but toward greater simplicity. In order for the evolution of complexity to occur, DNA has to be within a cellular context; the whole system evolves as a reproducing unit. So the notion of an autonomous replicator is another spot on the leopard that turns out to be an incorrect abstraction, and it fades out.

3. Weismann's dualism as a general biological principle is incorrect. All unicellular organisms, all plants, and many animal species, including mammals, have no separation of germ plasm from somatoplasm. The capacity to reproduce is a property of the whole *organism,* not a special replicating part that is distinct from the rest of the reproducing body. And in the case of sexual reproduction, to which Weismann's concept can be applied, it is the egg cell that carries the organization required for accurate replication of the DNA in the next generation, not a hereditary essence. Weismann's barrier persists in the idea that the DNA cannot be changed in an adaptive manner in response to environmental stimuli. According to current theory, an organism, when presented with a "problem" that it is initially unable to solve (such as using a new nutrient source), cannot introduce a "deliberate" change into its DNA to produce

a new protein that can make use of the nutrient. Such changes, if they occur, must arise in a random manner, and selection of those organisms with the randomly originating adaptive change can then follow.

However, this remnant of Weismann's theory, which is regarded as the most important aspect because it forbids Lamarckian inheritance, has now been challenged. There is new experimental evidence, initiated by John Cairns in 1988, that unicellular organisms such as bacteria and yeasts can indeed change their DNA in a directed, adaptive manner. The jury is out on whether Weismann's barrier is to remain in place or succumb to the versatility of the molecular mechanisms that make DNA a very fluid, rather than a stable, polymer. But as far as the reproductive process is concerned, organisms cannot be divided into a special hereditary essence that is potentially immortal and another part that is mortal. A *Paramecium* cell, for instance, divides to produce two similar cells, and each of these has the capacity to grow and divide, producing two more, and so on ad infinitum. Hence the old joke about the strange arithmetic of life: organisms multiply by division. As this process goes on, *all* the molecular components of the cell—the DNA, the RNA, the proteins—undergo what is called molecular turnover: their constituent parts get replaced, usually by identical subunits. They all get renewed, though the organism remains unchanged (i.e., the characteristics that make it a *Paramecium* rather than a *Salmonella* or a *Stenostomum*). In this respect organisms are like water fountains: the form remains the same but the material of which it is composed flows through it, changing continuously. What does not change in organisms are certain aspects of the *organization* of the materials—their dynamic relationships, the way they are arranged in space, and the patterns of change they undergo in time. Organisms are dynamic fields of particular kinds, as we shall see in more detail in the next chapter.

The hereditary material plays a very important role in stabilizing certain aspects of this spatial and temporal order. But it does not *generate* the order, and it is no more immortal than is the rest of the organization of the cell on which it depends for its replication. However, there is one very important respect in which the DNA is special and exceptional. It is the only macromolecular constituent of a cell that is accurately copied and partitioned to the two daughter cells during cell division. It is from this property that its importance as a stable transmitter of hereditary information derives. We saw in the last chapter that other aspects of cell structure can be accurately copied and transmitted to daughter cells, such as the reversed row of unit territories in the melon-stripe *Paramecium*. But this type of inheritance is restricted to species in which reproduction involves the copying of parental structures in producing the progeny. There are many species, both unicellular and multicellular, that reproduce this way. However, in sexual reproduction the only organization that is transmitted from parent to offspring is that carried by the egg cell, and the DNA of the chromosomes is the only molecular constituent that is precisely copied and transmitted in the process. Hence its importance. But if the DNA is not a generator of the structure of the organism that emerges from the egg, what is its role? How does it affect whether the new organism has two or four wings, five or six toes, whether it is tall or short? This is the question that will be considered in the next two chapters, where we look in detail at what happens during the development of an organism from an egg, and how the genes as well as other influences are involved in generating particular structures.

To summarize the emerging picture, it is useful to modify Figure 2.2 to take account of this evidence (Figure 2.4). Instead of germ plasm (or its modern version, DNA) as the carrier of the specific inherited factors from the parent that influence the formation of particular structures in the offspring, we have what I call *inherited particulars*. These can be either particular base sequences of DNA that define the genes, or particular structures of the parent organism, such as the melon stripe in *Paramecium,* that get transmitted. These act on the generative field, the organization of the egg cell or the organism itself that grows and develops to produce a new individual with characteristics inherited from the parents. Separating these two aspects of an organism is actually artificial, because in reality the two are part of a single system: the inherited particulars are part of the organism, which is a field that is organized in space and in time.

The brilliant light that shines on the genes as the most distinctive spot on the Darwinian leopard begins to fade into the context of the

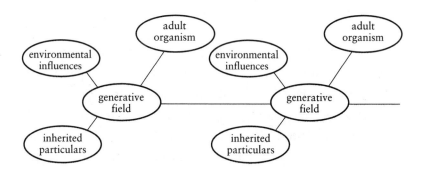

Figure 2.4   *Some of the components left out of the Weismann's scheme and not accounted for by molecular biology, in particular the organization of the generative field that produces the adult form with influences coming from DNA and cytoplasmic hereditary factors, and from the environment.*

whole organism, which now emerges more clearly as the living being. For certain purposes it remains useful to distinguish those aspects of this single dynamic field that are involved in transmitting particular influences from those that are responsible for generating form. This convention is actually used in the analysis of dynamic systems, as will become clearer in chapter 4, where a distinction is made between genes as parameters, and the generative field as the dynamical equations used to describe the process of morphogenesis. Distinctions of this kind are useful in analyzing complex systems, but they need to be recognized as abstractions. The organism itself is a unity.

We need two more parts to make the diagram in Figure 2.4 complete, as far as we have gone. First, there is the process whereby the generative field, influenced by inherited particulars, produces the adult form of the organism. This is the developmental process, and for the moment it is just the line shown from the generative field to the adult. Next, there is the role of the environment. So far I have said little about this, except to recognize that organisms require particular environmental conditions in order to survive. However, the environment can also have a specific influence on development by acting on the generative field, as shown. One of the most striking examples of this is sex determination in Mississippi alligators. It turns out that what determines whether a developing alligator will be male or female is the temperature it experiences during a critical period of its embryonic development. Eggs that develop in the temperature range 26°–30°C are all female, while between 34° and 36°C they are all male. In between 31° and 33°C, the switch occurs and eggs developing at these temperatures can be either sex, the probabilities changing from female to male as the temperature rises. Alligator eggs are not incubated at a constant temperature by the parent, as in birds, but are simply laid in a nest within rotting vegetation, which produces some heat. A clutch of twenty to thirty eggs experiences a range of temperatures, resulting in a mixed batch of females and males. Below 26°C and above 36°C the eggs fail to develop, so there are often undeveloped eggs in a nest. This species uses an environmental variable to regulate its sex ratio.

Most species do this via genes, as in humans with X and Y chromosomes—XX for a female, XY for a male. The result is a near fifty-fifty sex ratio, and this genetic method of determining sex seems to be a much more sensible way of regulating what is, after all, a very basic property of a population for survival. However, the alligator appears to do perfectly well, and has done so for a very long time. There are other reptilian species that use the same mechanism, such as lizards and turtles. So we certainly need to include environmental influences on the generative field in our diagram.

Within the diagram there are many other lines that can be drawn, describing different influences. The environment acts on the adult as well as on the developing organism. The adult can affect the environment: trees produce leaf humus, and their roots help to retain moisture in the soil; earthworms till the soil, which aerates it. The environment can influence the inherited particulars, as in the case of a *Paramecium* cell with the reversed row of cilia. There are also examples of the DNA being modified by the environment. But the focus of my inquiry now is on the nature of the generative field, which is what was left out of Weismann's scheme. It is the organized context within which inherited particulars act, and without which they can have no effect. Putting this back into biology leads to a new definition of an organism as the fundamental unit of life. And with this comes the need to look again at evolution. What will emerge is no longer a spotted leopard struggling for survival, but a rather different type of creature, described by different metaphors. However, the leopard will still be there if you want to see it.

Life, the Excitable Medium

It used to be believed that inside every human germ cell there is a little homunculus, a miniature human being, complete in every detail, that simply grows in the womb into a human infant. And so it was assumed for all other species. If you could see the miniature adult in sufficient detail, you would know what type of organism to expect from the egg of any species. The idea of a genetic program in an egg that specifies all the details of the organism by the information it contains is another version of this story. According to the program concept, everything we need to know to understand how an egg develops into an adult organism is written in its DNA. This is the old trap of assuming that whatever is created must result from the action of a creator whose word is written down in some form. But we have already seen in chapter 1 that a genetic program can do no more than specify when and where in the developing embryo particular proteins and other molecules are produced, and we know that molecular composition is not sufficient to explain physical form. Organisms are physical systems of a particular kind. The question is, what kind? This is what we need to know to explain how something as simple as a microscopic spherical egg can develop into an organism as complex as a baobab tree or an elephant. What kind of physical process is capable of generating this level of organized complexity, defining the dynamic context within which genes play an important, but limited, role?

Making Patterns from Nothing

All organisms have simple beginnings. Take the example of the common seaweed *Fucus,* found in coastal waters throughout the temperate oceans of the world. Tiny spherical eggs are shed from the adult fronds into the seawater, where they are fertilized by even tinier sperm. The eggs then slowly sink and attach to a rock with a sticky substance produced after fertilization. The first sign of development is the emergence of a little outgrowth from the bottom of the egg, so that the cell develops a pear shape (Figure 3.1). The egg has broken its initial spherical symmetry and developed an axis, the first step in forming a more complex shape. The outgrowth will develop into the rhizoid, the rootlike base of the alga, while the upper part will develop into the stalk and the leaflike frond that grows up toward the surface.

The formation of the axis is normally related to light. The egg, anchored to a rock by its sticky surface, has an upper illuminated part and a lower shaded part. The difference in illumination initiates axis

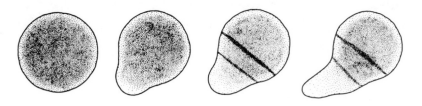

Figure 3.1   *The first stages of development in* Fucus: *breaking the symmetry of the spherical egg with a growing tip that initiates rhizoid formation and establishes the main axis of the alga.*

43

formation. Studies of this process in the laboratory reveal that there are many other stimuli that can initiate an outgrowth, for instance an electrical current or gradients of ions such as K^+, Cl^-, or Ca^{++}. And in the absence of any of these external influences, an isolated egg in a completely homogeneous environment will develop an axis anyway. There is something about the internal dynamic organization of an egg that makes spherical symmetry unstable, so that any perturbation, an internal fluctuation of ions, say, or an external stimulus, will get it started. What light and the other stimuli do is influence *where* the axis will appear, but it is not the cause of axis formation itself. It is just a trigger that initiates something that is poised and ready to go, like a sprinter at the start of a race. The situation is also like a previous example we considered: water flowing out of a bathtub. The organization of liquids is such that spiral flow is inevitable under these conditions; but whether the direction of motion is right- or left-handed depends on whatever influences happen to be acting that can initiate symmetry breaking and produce one of two possible stable spirals. In the case of the *Fucus* egg, the number of possible axes is infinite: an axis can form in any direction within the egg, regardless of what orientation the egg has when it lands on a rock. The axis will form along the gradient of illumination. The mechanism whereby this occurs depends on light-sensitive pigments in the egg that influence ion channels in the membrane, producing local flows of ions such as K^+, Cl^-, and Ca^{++} and so resulting in small electrical currents that can be measured. However, I want to focus not on how external influences act, but on what kind of dynamic organization in the egg gives it the irresistible tendency to break symmetry and to start on its path toward complex form. We turn to a fascinating bit of chemistry, which will give us the key.

Chemical Waves

Normally one thinks of chemistry as the art of making substances in test tubes. These may be colored; have a nasty smell; have a tendency to explode; or have useful properties such as dissolving stains, cleaning drains, or polymerizing to produce materials like plastic or synthetic rubber. What one usually does not associate with chemistry is dynamic patterns. However, in recent years a new dimension has been added to chemical reactions, and that is spatial order. When certain substances are mixed together in solution in a shallow, flat dish (a petri dish) and left undisturbed, lo and behold there emerge, spontaneously, beautiful and regular patterns. One of these magic mixtures is known as the Beloussov-Zhabotinsky reaction, named after two Soviet scientists who discovered and studied it in the 1950s and 1960s in Moscow. Their mixture of organic and inorganic chemicals generates concentric rings like a target pattern (Figure 3.2). The rings slowly travel outward from centers that arise spontaneously throughout the dish; new circles form at regular intervals. As can be seen, where the expanding rings encounter one another they disappear; they do not form interference patterns like waves caused by dropping a number of pebbles into a pond in different places. Each pattern retains its original form up to the boundary established by two colliding waves, which annihilate one another.

Here is something that in the 1950s was quite new in chemistry and extremely interesting. At first few paid any attention to the reports by Beloussov and his student Zhabotinsky; the phenomenon just did not fit into the mind-set of chemists, who were used to homogeneous reactions, not ones that generate patterns in time and space. But there

Figure 3.2   *The target pattern that forms spontaneously as waves propagate out from centers in a dish of Beloussov-Zhabotinsky reagent.*

were some who recognized the significance of this new dimension to chemistry, and, as so often happens in science, earlier reports of similar observations came to light. This is like the rediscovery of Mendel's laws of genetics decades after he had published them, albeit in an obscure journal. Ideas have their time, and if you happen to discover something before people are ready to recognize its significance, you might as well leave it in the bottom drawer until the climate is receptive. Beloussov's recognition did not come until after his death in 1970; in 1980 he was awarded the Lenin Prize. The whole story is told in *When Time Breaks Down* by Arthur Winfree, one of the first outside the Soviet Union to see the significance of the phenomenon and to get to work on it in the late 1960s.

Recognition of the importance of the Beloussov-Zhabotinsky reaction began to spread throughout the scientific community, and biologists in particular were fascinated when they realized that these purely chemical patterns were just like the patterns they had observed in a totally different system consisting of living cells. There is an extraordinary organism with the endearing name *cellular slime mold* that is ideal for studying the development of form. Its life cycle is divided into two very distinct phases. As long as food is available in the form of bacteria, this species exists as free-living, independent amoebas, single cells that crawl about engulfing and digesting bacteria. The cells go about their business of growing and dividing and pay not the slightest bit of attention to one another. But as soon as food runs out, they adopt a very different strategy. Cells start to signal to one another by means of a chemical that they release. This initiates a process of aggregation: the amoebas begin to move toward a center, defined by a cell that periodically gives off a burst of the chemical that diffuses away from the source and stimulates neighboring cells in two ways: (1) cells receiving the signal themselves release a burst of the same chemical; and (2) they move toward the origin of the signal.

A petri dish covered with these starving amoebas soon begins to develop spatial patterns remarkably like the expanding concentric circles of the Beloussov-Zhabotinsky reaction (Figure 3.3). What these

Figure 3.3   *Wave patterns in aggregating slime mold amoebas produced by movements of cells toward the source of the chemical signal.*

circles represent are waves of amoebas moving toward the signal, which is itself propagating out from the center, relayed by the amoebas themselves. The result is waves of cell movement toward a "founder cell" at the center that gives off a signal every five to eight minutes, many such centers arising spontaneously. Several thousand amoebas thus gather together and start to form a multicellular organism. They do this by passing through a sequence of stages shown in Figure 3.4, in which the initially simple aggregate of cells becomes progressively more complex in form, and the cells in different positions differentiate into specific cell types. The final structure consists of a base, a stalk that rises up from the base, and on top a "fruiting body" made up of a spherical mass of spores that can survive the absence of food and water. When conditions recur that allow growth, the spores are re-leased from the fruiting body and germinate—each one producing an amoeba that feeds, grows, and divides—and the life cycle starts again. So this species is ideal for studying the formation of a simple multi-

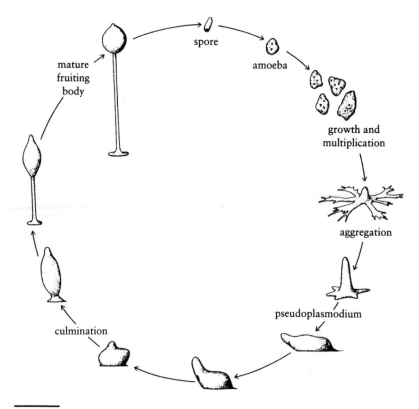

Figure 3.4   *The life cycle of the cellular slime mold* Dictyostelium
discoideum.

cellular structure and the differentiation of cells in different positions within the structure. But at the moment I want to focus on
the similarities between the pattern produced by a purely chemical
process and one arising from cell interactions. Are there basically
similar processes going on, or is this just an extraordinary coinci-
dence?

Different Molecules, Similar Patterns of Interaction

In terms of molecular composition, the Beloussov-Zhabotinsky reaction and slime mold aggregation have nothing at all in common. The
former is a complex cocktail of organic and inorganic ingredients

including such substances as malonic and bromic acids, and cesium ions. The signal in the case of the slime mold is closely related to the nucleotide bases that make up nucleic acids (RNA and DNA, mentioned in chapter 1). It is cyclic adenosine mononucleotide, or cAMP for short, and it is involved in basic energy regulation in all cells. In bacteria it is produced when cells are starving and is called the *glucose distress signal,* indicating that the cells have run out of their major energy supply, glucose. In the slime mold the same molecule serves the same function, but now it gets released in little bursts from starving cells. The signaling system consists of many components within the cells, such as enzymes involved in making cAMP, as well as an enzyme outside the cells that breaks it down. There are now models of these processes that describe the periodic signaling and the aggregation of the amoebas very effectively. There is also a detailed theory of the Beloussov-Zhabotinsky reaction and computer simulations that generate the target patterns. So both processes are now quite well understood, including the molecules involved and the basic properties of the reactions that result in spatial patterns. The two systems have quite different molecules; their similarities are in the relationships between the molecules.

Essentially what produces the patterns is a set of reactions with the following characteristics. First, there is a positive feedback effect so that a substance stimulates its own production. In the case of the Beloussov-Zhabotinsky mixture, the reaction is the following:

$$HBrO_2 + BrO_3^- + 3H^+ + 2Ce^{3+} \longrightarrow 2HBrO_2 + 2Ce^{4+} + H_2O$$

Bromous acid ($HBrO_2$), when mixed with bromate (BrO_3^-), acid (H^+), and cesium (Ce^{3+}) makes more bromous acid ($2HBrO_2$), oxidized cesium ions (Ce^{4+}), and water. Therefore, what happens in a dish of the reagent is that bromous acid starts to produce more of itself and the concentration builds up, stimulating the production of the acid in neighboring regions of the dish so that a wave of production spreads out. But then the second essential reaction comes into play: CO_2 is produced in the reaction and it inhibits $HBrO_2$ synthesis. There are two possible results of such a pair of reactions. Either the rate of production is just balanced by the rate of inhibition so that the system reaches a steady state and the net concentration remains constant, or production and inhibition seesaw out of balance and an oscillation occurs. Most chemical reactions reach a balanced condition, a steady state. The Beloussov-Zhabotinsky reaction is sufficiently complex that it oscillates. The result is that when the reagents are in an essentially two-dimensional space such as a thin layer of solution in a flat petri dish, propagating waves of bromous acid production are initiated from regions where the reaction gets a head start, often due to a bit of dust on whose surface the reactions are slightly accelerated, followed by a wave of inhibition.

In the cellular slime mold amoebas, cAMP is the molecule that stimulates its own increase, though the mechanism is quite different from that for bromous acid. An amoeba that is stimulated by cAMP releases it so that the concentration rises and the molecule diffuses into adjacent regions. Amoebas nearby are then stimulated by this diffusing cAMP to produce the signal, which then diffuses and stimulates other amoebas. So the signal propagates across the lawn of cells on a petri dish. But this is not enough to ensure an effective signal: it must also be destroyed, otherwise the whole dish of amoebas would become a sea of cAMP, and no signals would be visible. The amoebas secrete an enzyme, phosphodiesterase, that destroys cAMP. So the substance has a brief lifetime, and the diffusion profile of the signal from a stimulated amoeba has a steep gradient, generating an effective directional signal that allows other amoebas to use it for chemotaxis (directed movement in response to a chemical). However, there is a problem here: cAMP released from an amoeba diffuses symmetrically in all directions away from the source, so amoebas anywhere within the effective range of the signal could respond. This means that each stimulated amoeba could become the center of the propagating wave. The result would be total chaos. This does not happen, as is evident from Figure 3.3. The reason is beautifully simple and natural: after an amoeba has released a burst of cAMP, it cannot immediately respond to another signal and release another burst. It goes into a refractory state during which it is unresponsive, recovering from the previous stimulus and returning to its "excitable" condition. Therefore, the wave cannot travel backward, and the signal travels one way.

The refractory state accounts for another property of the aggregation process: the abrupt cessation of the signal wherever two waves collide. Since at such a point the amoebas just behind the wave front are refractory on both sides, the signal cannot propagate further, and the whole process stops. In the Beloussov-Zhabotinsky reaction, the refractory state is due to the fact that bromous acid is inhibited behind the wave front, so again the wave stops because the reaction cannot immediately switch to production again. In both cases an initially homogeneous system gets partitioned into separate domains, each under the influence of a center that arose spontaneously.

These examples show that what counts in the production of spatial patterns is not the nature of the molecules and other components involved, such as cells, but the way these interact with one another in time (their kinetics) and in space (their relational order—how the state of one region depends on the state of neighboring regions). These two properties together define a field, the behavior of a dynamic system that is extended in space—which describes most real systems. This is why fields are so fundamental in physics. But a new dimension to fields is emerging from the study of chemical systems such as the Beloussov-Zhabotinsky reaction and the similarity of its spatial patterns to those of living systems. This is the emphasis on self-organization, the capacity of these fields to generate patterns spontaneously without any specific instructions telling them what to do, as in a genetic program. These systems produce something out of nothing. Now, we can see precisely what is meant by "nothing" in this context. There is no plan, no blueprint, no instructions about the pattern that emerges. What exists in the field is a set of relationships among the components of the system such that the dynamically stable state into which it goes naturally—what mathematicians call the generic (typical) state of the field—has spatial and temporal pattern. Fields of the type we have been considering are now called *excitable media*. In the rest of this chapter, we see evidence from a diversity of examples showing that many of the properties of organisms and their parts can be understood as the dynamic properties of excitable media. Then, in the next chapter, we will look at how these ideas can be applied to morphogenesis, how organisms generate their forms.

A Zoo of Patterns

The target patterns described earlier are not the only ones that emerge in excitable media. Another one that made its appearance in both the Beloussov-Zhabotinsky reaction and slime mold aggregation is a spiral form, unwinding from a center like a slowly revolving Catherine Wheel. These can be seen in Figure 3.3 along with the target patterns, as they form spontaneously in aggregating amoebas. To get spirals in the Beloussov-Zhabotinsky reaction, it is necessary to disturb the system by tipping the petri dish gently to one side, introducing a slight shear. This initiates spirals, as in Figure 3.5. Once they get started, the spirals tend to take over, displacing the concentric circles. The reason for this is connected with the observation that the wavelength of the spiral pattern (the distance between the center of one arm of the spiral and that of its neighbor) is slightly smaller than that of a target pattern, which means that the periodicity of the generating center of a spiral is slightly smaller than that of a target pattern. So a

Figure 3.5   *Spiral waves in the Beloussov-Zhabotinsky reaction.*

spiral is a bit quicker than a concentric circle in getting the next wave started. It turns out that this is because the spiral initiator is a wave propagating around in a little circle at the center with a cycling time that is just equal to the refractory period of the amoebas and the Beloussov-Zhabotinsky reaction. This is the *minimum* time that there can be between initiations. The target pattern initiator, on the other hand, is usually a little slower than this, because it depends on the cycling time of the periodic process in repetitive signal release from an amoeba or in the Beloussov-Zhabotinsky reaction, which drifts about a mean value. What happens when spirals get started is a take-over of the whole field by these faster "replicating" patterns, like the naked replicators in the test tube that Spiegelman studied (chapter 2). This type of displacement of one type of system by another is often described as an example of natural selection in a test tube. What this makes clear is that *there is nothing particularly biological about natural selection:* it is simply a term used by biologists to describe the way in which one form replaces another as a result of their different dynamic properties. This is just a way of talking about dynamic stability, a concept used for a long time in physics and chemistry. We could, if we wished, simply replace the term *natural selection* with *dynamic stabilization,* the emergence of the stable states in a dynamic system. This might avoid some confusion over what is implied by natural selection.

There are other patterns that can arise in excitable media. Instead of waves propagating from centers, the whole system can change periodically from one state to another so that changes occur in time, but they are spatially homogeneous. This happens in the Beloussov-Zhabotinsky reaction if the reagents are kept well mixed by stirring, and similarly in amoebas mixed together in suspension. The intrinsic periodicities of the reactions are then revealed. Other patterns can occur if the concentrations of the reagents are changed sufficiently to abolish this intrinsic periodicity. For instance, a wave can be started at one point by an external stimulus, and this can propagate over the whole system as a single widening circle that stops at the boundaries, after which the system goes quiet unless stimulated again.

One of the most interesting patterns that has emerged was one that looked completely disorganized, with no coherence at all in the waves. However, a closer examination of this by Harold Swinney in Texas and his two French collaborators, J-C. Roux and Reuben Simoyi in Bordeaux, revealed that this was an example of deterministic chaos. Figure 3.6 shows the pattern of change of one of the chemicals (the bromide ion) in both a reaction with regular periodic activity and a reaction with irregular activity. The power spectra reveal the frequencies, showing the regularity of one and the irregularity of the other. However, there is a way of analyzing dynamical systems that provides important insights into their behavior. This involves looking at the relationship between the state of the system at time t and comparing it with the state at a later time, $t + T$, where T is chosen according to the rate of the reaction. Since the Beloussov-Zhabotinsky reaction goes through a cycle every few minutes, it is necessary to choose a value of T such that many states are compared in every cycle, to give a picture of the dynamics. Figure 3.7 (right panel) shows the result of such an analysis, with $T = 0.88$ secs., which was the time interval used between measurements of bromide concentration in the experimental apparatus. What this figure reveals is that the system is moving on orbits (motion from left to right in the lower part of the figure) such that they have some order, remaining in a well-defined region even though they are not periodic. The region of the trajectories is called an *attractor,* and in this case the attractor is called *strange* because of the way the orbits wander in an unpredictable way, while remaining confined to the region, characteristic of a system in a state of deterministic chaos. By contrast, when the Beloussov-Zhabotinsky reaction is in periodic mode, the attractor is a closed curve representing

Figure 3.6   *Comparison of the Beloussov-Zhabotinsky reaction in two different dynamic modes, one with a regular periodic pattern (top) and one without (bottom). The power spectra are shown in (b), revealing no major frequencies in the irregular pattern, in contrast with the periodic system.*

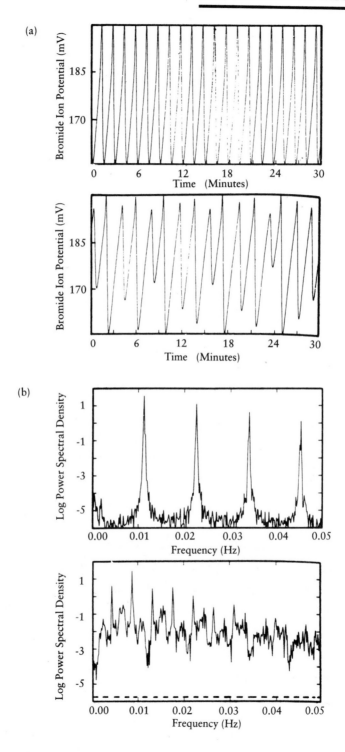

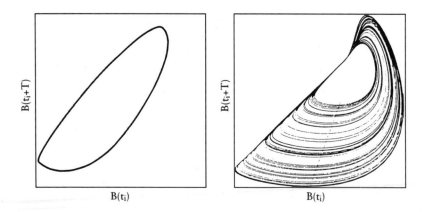

Figure 3.7   *The structure of the strange attractor for the chaotic mode of the Beloussov-Zhabotinsky reaction (right), compared with a typical attractor for a periodic system (left).*

a strictly repeating orbit (Figure 3.7, left panel). (A system that is random in its activity would show a dense pattern of orbits that completely fills up a region.) Compared with a purely random process, deterministic chaos has a lot of order, but it does not show up unless one looks at the system in a particular way. Characterizing dynamical systems in terms of their attractors—the pattern of their orbits—is a very useful way of describing their qualitative behavior.

Beautiful patterns arise when the dish of Beloussov-Zhabotinsky reagent is deepened so that the system has a third dimension in which to express itself. Then what can be seen are scroll waves, spirals that are like sheets wrapped in a scroll, though in this case the sheets are not stationary but are propagating waves. Within this added dimension it is possible to imagine other possibilities, such as twisted scroll waves, in which the generator at the center is not a wave traveling in a circle but is traveling around as if on the surface of a central doughnut or torus. This and other exotic patterns have been proposed by Arthur Winfree and his colleague Steven Strogatz, and simulated in three dimensions on computer. It is not easy to initiate such waves in a Beloussov-Zhabotinsky reaction because they require special initial conditions to get them started. However, the slime mold does produce

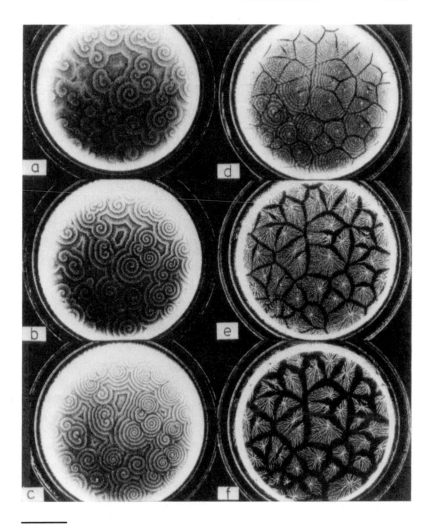

Figure 3.8   *The gradual breakup of the original lawn of amoebas into separate aggregates around initiating centers. Each aggregate forms an organism of several thousand cells, which then develops as shown in Figure 3.4*

one of these exotic patterns, but only when it has itself become a three-dimensional organism.

The original lawn of amoebas gradually breaks up into a set of separate aggregating domains, each centered on a signal source, as shown in Figure 3.8. The result is a partitioning into roughly equal-sized aggregates of several thousand cells each. Each aggregate forms

a mound of cells that piles up and then topples over, forming a multicellular slug that migrates over the surface leaving a slime trail behind
it, hence its name, the cellular slime mold (see Figure 3.4). As the slug
moves, the cells differentiate according to their position: cells at the
anterior end begin to change along a pathway that ends up as the stalk
of the fruiting body, while the majority of the cells, posterior to the
tip, differentiate into the spore cells that end up in the sorocarp of the
fruiting body. Periodic propagating waves continue to play a role in
the dynamic organization of these processes of slug migration and cell
differentiation. The waves are initiated at the tip of the slug and propagate toward the rear, cells moving forward within the slug with a
periodic change of velocity as they respond to and transmit pulses of
cAMP, the frequency being about one wave every two minutes.

Siegert and Weijer in Munich have shown that the generator of the
periodic waves is, once again, a rotating wave at the tip. However,
this is now initiating a wave in a three-dimensional structure, so the
geometry of the wave is somewhat more complicated than the spirals

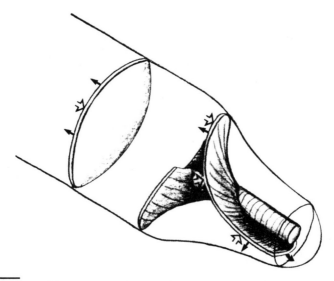

Figure 3.9   *The structure of a scroll wave at the tip of a slime mold slug,
showing how the scroll unwinds to a planar wave that propagates through the slug from front to back, while the cells
move forward, propelling the slug by their collective motion.*

in a plane that occur during aggregation when the cells are spread out on the agar surface. What Siegert and Weijer have deduced from the detailed observations of individual cell movements and the wave pattern is the three-dimensional waveform shown in Figure 3.9. Looking at the tip, you would see a rotary wave propagating in the direction of the arrow on the circle, with cells moving in the opposite direction, toward the signal source. So cells rotate around the tip at right angles to the forward movement of the slug. But as the wave travels back along the slug, it changes its shape and transforms into a planar wave that propagates backward through the slug, with cells moving forward and giving coordinated movements to the whole.

A Very Excitable Organ: The Heart

What is a heartbeat? It is a propagating contraction wave that travels from an initiation center (a pacemaker) called the sinoatrial node (SAN) in the right atrium of the heart (where the blood flows in from the body), travels through the muscular tissue of right and left atria, accompanied by muscular contraction, and arrives at the atrioventricular node (AVN). Specialized conducting fibers (Purkinje fibers) ramifying from the AVN to the muscle tissue of the ventricles (VM) initiate the main pumping action of the heart in these chambers, a strong muscular contraction, accounting for 80 percent of the power in each stroke (see Figure 3.10). The tissues then recover in preparation for the next cycle. This recovery includes the restoration of an electrical potential difference across the membranes of the conducting cells and the muscle cells, which are about 90 millivolts more negative inside than outside, just like nerve cells. This involves the action of pumps in the membranes that remove potassium (K^+) from inside the cell and allow sodium (Na^+) in. During this recovery, which lasts for 100–200 milliseconds, depending on the heart rate, the cells are refractory and will not respond to another electrical stimulus. This ensures that the heart does not backfire and contract in the wrong direction, for the same reason that in aggregating slime mold cells the signal does not reverse

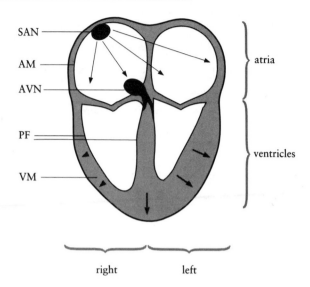

Figure 3.10   *The electrical conduction system of the heart from the pace-maker in the sinoatrial node (SAN) through the atrial muscle (AM) to the atrioventricular node (AVN) and to the ventricular muscle (VM) via the Purkinje fibers (PF).*

its direction of propagation and destroy the coherence of the process.

The heart can function on its own, without any external nerve supply, though of course there are major nerves that influence its activity according to the condition of the rest of the body. It is a self-exciting system, designed to operate in a particular dynamic mode—the familiar thump-thump-thump of the repeating contraction wave pumping blood around the body. Because the signal that propagates over the system is primarily electrical with a short recovery time, both the velocity of the wave and the frequency of the rhythm are much greater than the waves in the Beloussov-Zhabotinsky reaction and the aggregating amoebas. But otherwise the processes are remarkably similar. We are dealing with a third example of an excitable medium that can show the same range of dynamical behavior as those that we have encountered in the other two. One of these dynamic modes turns out to be bad news for heart function, as we shall now see.

Despite all the backups that are built into the heart to keep it functioning in its coherent contraction mode, initiated by the pacemaker in the sinoatrial node and terminating with the powerful pumping action of the ventricles, a healthy, normal heart can fail in the most dramatic and unexpected way. The phenomenon of sudden death due to cardiac failure claims 100,000 victims per year in the United States alone. While postmortem autopsies often reveal that this is due to slight damage to heart tissue, often caused by ischemia (inadequate blood supply to the heart tissue itself due to some circulatory occlusion such as a clot or thickening of the arteries), there are many instances in which the heart tissue appears to be perfectly healthy. It seems that there is some kind of dynamic black hole that the heart can fall into, one of the normal states available to it that is incompatible with effective pumping action.

Sudden cardiac arrest is not, in fact, due to the heart stopping its activity. Over a century ago, in 1888, the cardiologist who named the phenomenon, J. A. MacWilliam, described it as dynamic disorganization rather than cessation: "The cardiac pump is thrown out of gear, and the last of its vital energy is dissipated in a violent and prolonged turmoil of fruitless activity in the ventricular wall." This condition of uncoordinated contraction is now called *fibrillation*, a kind of repetitive quivering of the heart muscle as waves of ineffective contraction pass over it, initiated from sites in the ventricle itself. An example of this is shown in Figure 3.11. The normal heart beat at about one-second intervals was suddenly interrupted by a premature ventricular contraction that initiated the fibrillation with a frequency of about one contraction every 200 milliseconds (five beats per second). This then degenerated into an incoherent pattern. Fibrillation can be observed clinically in the atria, where it is not lethal, but once the ventricles start to fibrillate the body is in danger unless the condition is quickly corrected. What initiates this condition?

Cardiologists working in Maastricht, the Netherlands, reported the following case in 1972. A fourteen-year-old girl suddenly lost consciousness when awakened one night by a thunderclap. This suggested anoxia due to circulatory failure, but the condition was transient and the girl regained consciousness without any ill effects. However,

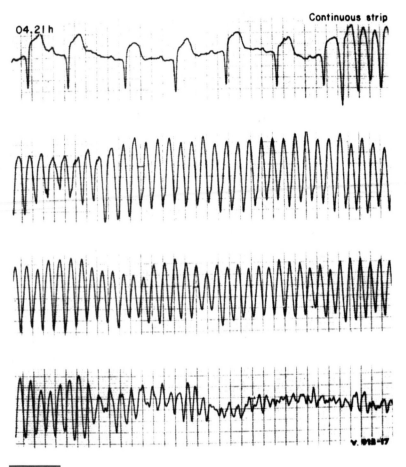

Figure 3.11   *The electrocardiogram of a patient who was wearing a recorder at the time of onset of the fibrillation that led to heart failure.*

she would then often faint when startled by her alarm clock in the morning, recovering a few minutes later. The cardiologists established that this was indeed due to ventricular fibrillation.

A common cause of such behavior is the stimulation of the heart from the autonomic nervous system, which modulates and coordinates heart activity. Autonomic fibers ramify across the heart tissue, influencing it by release of specialized molecules such as norepinephrine that depolarize cardiac membranes via receptors in the membrane. These substances are known as *beta blockers,* and their action can be stopped by an artificial beta-blocker molecule such as propanolol that sits on the receptor and prevents the normal molecules from acting. So the girl was given this, and her condition improved. After a few years she was persuaded by a friend to discontinue the drug. She was found dead in bed fourteen days later. It seems that the excessive sensitivity of her heart to autonomic nerve stimulation resulted in the onset of a lethal fibrillation. There was nothing wrong with her heart itself. Some property of excitable tissue makes it prone to this dynamic pattern.

There is now good evidence what this property is likely to be. Experimentalists have shown that tissue taken from the heart of a sheep and kept in physiological salt solution with glucose and plenty of oxygen can be stimulated with electrodes in such a way that it fibrillates. And the fibrillation is due to a dynamic pattern that is already familiar: propagating spirals initiated by circular rotation of an excitation wave at the center of the spiral, just as spirals are produced in the Beloussov-Zhabotinsky reaction and the aggregating slime mold amoebas. Remember that the centers of these wave patterns are rotating with a period that is close to the refractory time of the system, so that there is just time for recovery from one excitation before the circular wave closes the loop and another excitation occurs. For heart tissue the refractory period is 100–200 milliseconds, and these are indeed the periods of repeated contractions observed in fibrillating hearts, as in Figure 3.11 where the interval between contractions during fibrillation is about 200 milliseconds.

This hypothesis about the dynamic origins of fibrillation explains why small areas of damage to heart tissue, called *infarcts,* caused by temporary reduction of blood supply to the tissue due to a small clot or thickened arteries, are dangerous even though they may occupy insignificant areas of the ventricle. When the normal excitation wave of a heart contraction arrives at such an island of damaged, unexcitable tissue, it has to travel around it. The result can be that the wave then begins to rotate about the infarct, setting up a rotor that acts as an initiator of repeating contraction waves so that the ventricle goes into fibrillation. The infarct, insignificant in itself, acts as a facilitator of a natural dynamic mode that is incompatible with coherent pumping activity of the heart and so can be fatal. However, the demonstration that fibrillation can occur in perfectly healthy tissue provides good evidence for the hypothesis that sudden cardiac death is to be regarded as essentially a dynamic disease in which an excitable medium accidentally switches from one mode to another. Whereas for aggregating slime mold amoebas it does not matter much whether aggregation is centered on a periodic pacemaker or a rotor giving spiral waves, and a switch from one to the other can happen, in the heart the consequences are much more dramatic. But both are expressing the same properties of excitable media, giving us an integrated perspective on dynamic pattern formation in biological systems that at first sight seem to have nothing in common at all.

Brain Waves

An obvious candidate for excitable activity is the brain. In fact, the whole theory of excitable media really began with studies of the types of dynamic activity that can occur in a tissue made up of interacting neurons. The electrical signal that travels along a nerve cell is the same as that which activates heart tissue, with a flow of ions (Na^+ and K^+) across the membrane during the depolarization phase and then a recovery phase during which the nerve is refractory and cannot be stimulated again. A population of neurons that are connected to one another by little processes ramifying out from the cell body, as occurs in the brain, makes up a network that behaves as a continuous excitable medium. This will conduct activity waves one-way across the tissue from an initiating center. A remarkable technique called magnetic resonance imaging (MRI) has been developed in recent years for looking into living tissues without in any way disturbing their function. Extremely sensitive magnetic field detectors called SQUIDS (superconducting quantum interference devices), responsive to single quanta of magnetic field flux, are used to measure the magnetic fields produced by electrical currents flowing in, for instance, the brain. A sophisticated solution to what is called the *inverse problem* (deducing the electrical current distribution that produced the magnetic field), developed by A. Ioannides at the Open University in the United Kingdom and programmed to convert the observed magnetic field into the original electrical current distributions in the tissue, produces the type of image shown in the upper half of Figure 3.12. It can be seen that activity waves are initiated periodically from the thalamus and propagate to the cortex.

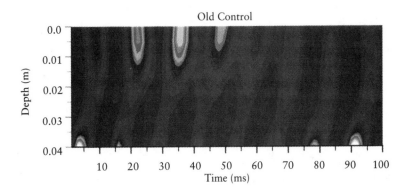

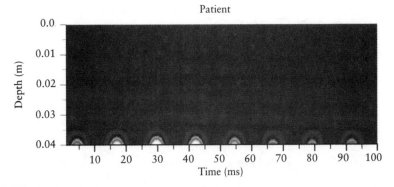

Figure 3.12   *Coherent waves of electrical activity propagating from the thalamus, deep in the brain (depth = 4 cm.), to the cortical (surface) layer, in a normal seventy-four-year-old (Old Control). In an Alzheimer's patient the pacemaker in the thalamus is normal, but the coherence is much reduced and no propagating wave is seen. (The method of representing these waves of electrical activity actually doubles the real frequency, so the pacemaker is running at about 40 hertz, or cycles per second, or with a periodicity of 25 msec. rather than the 12.5 msec. shown.)*

In a normal, healthy individual (Old Control, age seventy-four) these waves are coherent and appear to play an important role in coordinating the functions and activities of the cortex. In an Alzheimer's patient, on the other hand, this coherence is lost, as shown in the lower image of Figure 3.12. The pacemaker in the thalamus continues to function normally, but the coherence of the propagating waves is lost and no periodic pattern of activity reaches the cortex. This points to some aspect of the dynamic organization of brain tissue that is lost and results in the functional failures associated with this degenerative disease. Noninvasive techniques such as MRI, which can reveal the dynamic activities of the body without interfering with them, are extremely important in developing insight into the purely dynamic aspects of disease states. They put us on the threshold of a new understanding of how large-scale, coordinated activity in living systems emerges from the properties of components when they are organized and interact in particular ways. There are constant surprises arising from this work, and one of the most remarkable is how closely connected order and chaos are.

Odor and Chaos

A rabbit is being trained to recognize certain rather pleasant smells, such as those of pears and bananas. The rabbit is thirsty, and if it licks in response to a particular odor it gets water as a reward. With a different odor, it gets a drink if it simply sniffs and does not lick. So the rabbit learns to distinguish between the two odors and respond in particular ways: its behavior becomes conditioned by the stimuli. This is a classical conditioning procedure, first used by the great Russian physiologist Ivan Pavlov. While the rabbit is learning to discriminate between the different odors and to associate these with different behavior patterns, the electrical activity is being monitored by electrodes in the olfactory bulb of its brain, which responds to odors transmitted to it from sense receptors in the nose. The resulting record, an electroencephalogram (EEG), can then be analyzed to try to identify what specific changes in activity correspond to the process of learning to discriminate different odors.

This type of study has been going on for many years, but discovering the relevant signals in the noise of EEG patterns has been remarkably difficult. In recent years, combinations of new ideas about the dynamics of complex systems like the brain and sophisticated techniques of electrical recording, filtering, and analyzing signals have resulted in novel insights into what may be happening in neural tissue during learning. One has come from Walter Freeman and his group at Berkeley in collaboration with Christine Skarda and her colleagues at the Ecole Polytechnique in Paris, who conducted a detailed study on rabbits learning to discriminate odors by the method just described. Here is what they found.

The EEG of the olfactory bulb of a rabbit that is awake but not aroused by any novel or interesting odors looks like random noise, but is in fact best described by the state of deterministic chaos. This state is indistinguishable from noise in its appearance and in its statistical properties, but it is produced by a perfectly deterministic process and is not stochastic (random); it has a particular kind of order, like that shown in Figure 3.7 for the chaotic condition of the Beloussov-Zhabotinsky reaction. Noise is, in fact, very difficult to generate in a dynamic system such as an excitable medium that is also capable of producing ordered patterns of the type we have been considering, but chaos can readily be generated. This seems to be what is happening in the olfactory bulb, because when an odor is presented to a rabbit, there is a change in the pattern of activity from chaos to bursts of oscillatory activity with periods of about 20 milliseconds. The amplitudes of these oscillations varied in a systematic and repeatable pattern across the olfactory bulb at a common frequency. There was no evidence that odor recognition was localized to particular information channels in the olfactory bulb. Rather, it was the global pattern over the whole bulb that appeared to define the signal that was transmitted to higher levels of the brain for further processing. This distinctive spatial pattern was produced repeatably when a rabbit inhaled air that contained a particular odor. When the rabbit exhaled, the pattern over the olfactory bulb disappeared and was replaced by chaos. So learning to respond to an odor was accompanied by a transition from chaos to antichaos (order) and back again.

Skarda and Freeman used these and other observations to construct a model of brain activity that has a number of distinctive features, and explains the experimental results rather well. A basic property of the model is that the system simulating the activity of nerves in the olfactory bulb undergoes sudden changes of state from chaos to rhythmic activity and back again as an excitatory stimulus is increased and decreased, similar to the state transitions observed in the EEG of the olfactory bulb as the rabbit inhales an odorant stimulus and then exhales, returning to the chaotic rest state.

There is clearly a lot more to the rabbit's brain than production of patterns of activity in response to particular sensory stimuli. This is just the beginning of a complex process in which neural activity patterns at one level get transmitted to higher centers, where integration with other signals results in signal transmission to muscles for coordinated responses such as licking and drinking, which then activates return circuits registering thirst, and so on. However, getting some insight into the basic dynamics of the underlying activities in neural networks is obviously crucially important. And what we see in the work of Skarda and Freeman is further evidence that neural tissue behaves as an excitable medium with characteristic dynamic modes, including chaos. In the next section we see the significance of chaotic dynamics and its relation to pattern in a system whose emergent properties have often been compared to those of the brain: an ant colony.

Consider the Ant and Be Wise

Ants and other social insects, such as termites and bees, present us with a paradox that is good for proverbs. The activity patterns of individuals often appear to be very disorganized, and by any of the normal criteria they certainly cannot be described as intelligent. No one has succeeded in teaching individual ants anything; for instance, they are simply incapable of learning to discriminate one direction from another in finding a food source, always making the same random choice at a Y junction even if the food is always in the same place. However, put a bunch of ants together, and what marvels of collective activity result! Individual neurons are not very intelligent either, but a lot of them hooked up together can result in remarkably interesting and quite unexpected behavior. This is the very essence of emergent behavior. For biologists interested in such matters, ants and colonies offer an advantage over neurons and brains in that they are easier to observe.

Ant species of the genus *Leptothorax* are very convenient for laboratory study. Colonies typically consist of a hundred or so individuals, and the ants are small so that in their natural habitat they make their nests in hollow acorns or narrow crevices in rocks. In the laboratory they will happily make a nest in the space provided by a slightly elevated cover glass on a microscope slide, with a source of sugar solution nearby. Under these conditions they can be easily examined. Blaine Cole, working in Houston, Texas, made videotapes of colonies consisting of different numbers of ants and then analyzed their movements. When there were only a few ants in the space provided, each individual moved in a chaotic pattern, characteristically moving about for a period of time and then stopping and remaining immobile for a while before moving again. Both their bouts of activity and their periods of rest appear to be of random duration, but actually fall on a chaotic attractor.

As the density of the colony was increased by adding more ants to the defined territory, Cole observed a sudden transition to dynamic order: patterns of activity and rest over the colony as a whole (measured as the sum of numbers of individual ants moving or resting) suddenly changed from chaotic to rhythmic with an average period of about 25 minutes. The colony is taking a break a couple of times every hour! The difference between these activity patterns is shown in Figure 3.13. Rhythmic activity in colonies is measured by the number of grid units in the territory over which ants move. Fourier transforms of this data, which reveal any prominent rhythms as frequencies, show well-defined frequency peaks. The activity is still quite "noisy," for some individual ants continue to be active when the majority are at rest, and some are at rest during activity bouts of the colony as a whole. In contrast, among individuals or small colonies, the absence

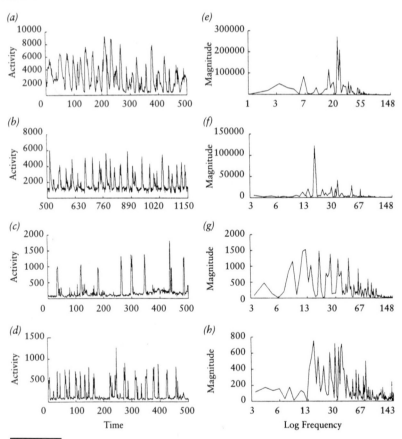

Figure 3.13   *Rhythmic activity patterns in typical ant colonies, (a) and (b), with the associated power spectra or Fourier transforms shown in (e) and (f), with clearly defined peaks at about 25 mins. In contrast, individuals either isolated or in a low-density colony, (c) and (d), have no well-defined frequency, (g) and (h), and are in a state of deterministic chaos.*

of any rhythmic components in the activity pattern is revealed by a Fourier analysis from which no single frequency emerges. The transition from chaotic movement in isolated individuals or low-density colonies to rhythmic behavior in a regular colony occurs when the density reaches a critical value: the group behaves in a collective mode that cannot be predicted from the behavior of individuals. This is a clear example of emergent behavior. One way of trying to understand how this occurs is to make a model and see what happens.

Modeling an Ant Colony

Why should density of ants play an apparently crucial role in the transition from chaotic to ordered behavior? Ants interact with one another, and an active ant encountering an inactive one will stimulate the latter into movement. At low densities there are few encounters, but at higher densities activity can spread like a contagion through a colony. This seems to be the simple key to the process, and it is just like the spread of activity in an excitable medium. Now ants become the excitable elements.

Using this approach, Octavio Miramontes, from Mexico, working with me at the Open University in collaboration with Ricard Solé, from Spain, constructed a model of ant colony behavior that provided intriguing insights into the dynamics of this system. Individual ants were modeled as elements driven by chaotic dynamics, represented by a neural network in the brain functioning in a chaotic mode and controlling movement. When in an active state, individuals moved from one site to any adjacent, unoccupied site on a lattice that defined the territory of the colony (Figure 3.14). If an individual moves to a site adjacent to one occupied by an inactive ant, the latter is stimulated to become active, but its activity will cease after a time unless it becomes spontaneously active or is stimulated by another ant. The whole model is rather like a neural network except that the elements can move. Because the elements (ants) obey simple dynamic rules, they are called *cellular automata,* and in this case they are mobile cellular

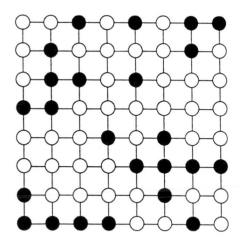

Figure 3.14   *Grid representing the territory of a colony of ants, which are represented as cellular automata that move on the lattice according to simple rules. Black circles are occupied sites; white ones are unoccupied.*

automata. These are very useful in studying the properties of complex dynamic systems of various kinds, made up of interacting elements. They have been widely used in the study of artificial life, about which I have more to say in chapter 6.

In the model, colonies of one or a few individuals had chaotic movement patterns, as shown in Figure 3.15, with the Fourier spectrum showing a broad range of frequencies (Figure 3.16). However, above a critical density there was a transition to rhythmic activity over the colony. At higher densities the rhythm becomes very pronounced and regular, but this is beyond the range of densities normally encountered in real colonies.

So here is another example of an excitable medium in which there is the same type of transition from chaos to order (and back, as density varies) that Skarda and Freeman deduced from their observations on the olfactory bulb of rabbits, supporting the view that neural networks and ant colonies indeed reveal similar types of emergent behavior. Now it becomes clearer why. Both are examples of excitable media in which the units, though quite different from one another, interact in

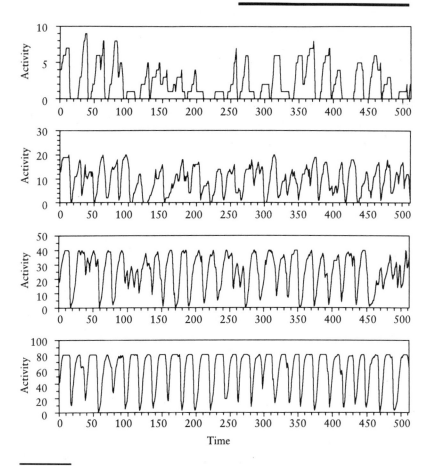

Figure 3.15 *At low densities (upper sequence, density 0.1) the activity pattern is chaotic, but as the density is increased a colony-wide periodicity emerges (densities 0.2, 0.4, 0.8 in successive sequences).*

similar ways so that waves of activity pass over the system. These waves can be either chaotic or ordered, depending on the state of the system. In the case of ants, density controls the transition; in the olfactory bulb, some activation threshold of the neurons varies with the state of attention and arousal of the rabbit, rather like the modulation of excitability of heart cells that results from the action of the vagus nerve and that can initiate rhythmic activity—ventricular fibrillation. These are all dynamic modes of excitable media and transitions between the modes.

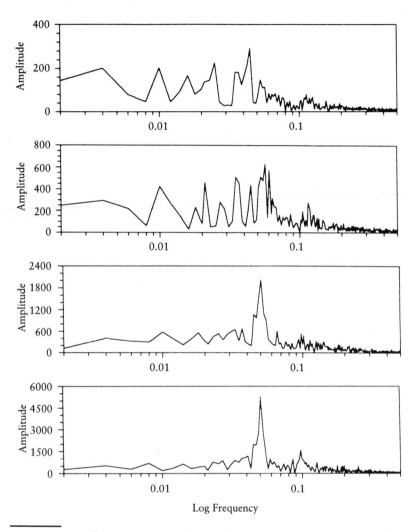

Figure 3.16   *The Fourier transforms of the activity patterns in Figure 3.15, showing the progressive emergence of a well-defined frequency as the density increases.*

Remembering where we started this inquiry into excitable media—with the spatial patterns of the Beloussov-Zhabotinsky reaction and aggregating slime mold amoebas—it is to be expected that ant colonies would develop spatial order at the same time as a coherent rhythm emerges. The model shows that this is indeed what happens, as can be seen in Figure 3.17. Concentric circles of activity develop, with

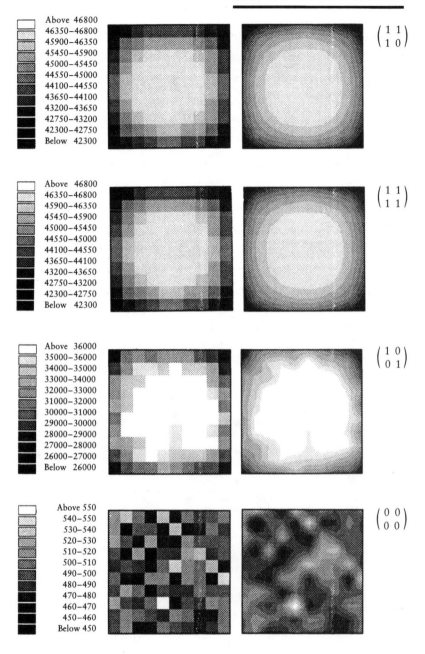

Figure 3.17   *Spatial patterns of activity in colonies, with activity increasing toward the center in colonies with interacting ants (upper three rows, different patterns of interaction) but no activity when ants do not interact (bottom row). The patterns on the right are time-averaged versions of those on the left.*

75

ants becoming more active toward the center of the colony and less active toward its boundaries. The nest or brood chamber of an ant colony is also organized spatially in concentric circles, but not according to activity. The queen, if there is one, is located at the center of the brood chamber with the developing ant embryos and larvae (the brood) arranged in concentric circles according to age, with the oldest nearest the boundary of the nest. The model certainly does not explain this organization; it simply shows that spatial order arises with temporal order, and the ants then make use of these emergent properties in whatever ways are appropriate to their activities.

We are now in a position to use ideas about the behavior of excitable media to see how organisms generate their forms, the question we started with in this chapter. This is the content of chapter 4.

Living Form in the Making

The last chapter provided illustrations of a rather remarkable phenomenon: similar patterns of activity can arise in systems that differ greatly from one another in their composition and in the nature of their parts. It does not seem to matter much whether we are dealing with chemical reactions, aggregating slime mold amoebas, heart cells, neurons, or ants in a colony. They all show similar types of dynamic activity—rhythms, waves that propagate in concentric circles or spirals that annihilate when they collide, and chaotic behavior. The important properties of these complex systems are found less in what they are made of than in the way the parts are related to one another and the dynamic organization of the whole—their relational order. The detailed behavior of the system is clearly dependent in important respects on the properties of the processes involved (for instance, whether they are chemical or electrical), since these determine rates of change and hence propagation velocity of the waves and frequencies of rhythms. But the patterns cannot be predicted from a knowledge of the properties of the component parts in isolation. To understand these complex nonlinear dynamic systems it is necessary to study both the whole and its parts, and to be prepared for surprises due to the emergence of unexpected behavior such as rhythmic activity, whether from chaotic neurons or ants. In this sense the study of complex systems goes beyond reductionism, which focuses on the analysis of the components out of which a system is made. This works well for simple mechanisms. It is of much more limited value for organisms. Now I am going to consider how we can study and understand organisms as dynamic wholes without losing any of the insights that have come from the investigation of their parts. This will lead toward an understanding of why life is capable of such diversity and beauty of forms while at the same time revealing an underlying unity. These are big questions in biology, but they have their roots in the simplest of organisms.

Acetabularia acetabulum is a little organism with a big name. It is an alga that lives in shallow waters around the shores of the Mediterranean. Look out for it if you ever take a trip to one of the southern Greek islands such as Rhodes. Attached to rocks you will see clumps of these green algae with stalks one or two inches long and beautiful little "parasols" or caps dancing with the movement of the waves (Figure 4.1). It is these that give the species its common name of Mermaid's Cap. Although it appears to be made of many cells, particularly because of the detailed structure of the cap, it is in fact all one single cell whose nucleus lives in one of the branches of the rootlike structure at the base (the holdfast or rhizoid). So this is a giant unicellular green alga, a member of a group called the *Dasycladales* that has lived in warm seas around the world for nearly 600 million years. They used to be much more numerous; now there are only twenty or so surviving species. However, they have had a good run, and there is no reason why they should not continue for many millions of years unless we succeed in polluting all the oceans.

Organisms perpetuate themselves by what is probably the most basic of all biological phenomena, a life cycle. Parts of the adult create a new whole organism by a process in which an initially simple struc-

Figure 4.1   *A colony of* Acetabularia acetabulum *attached to a rock in its natural habitat, the Mediterranean.*

ture develops into the adult form, which then produces the parts that will start the next generation. In *Acetabularia* this cycle (Figure 4.2) starts with the fusion of two tiny motile, flagellated cells called *isogametes*. These reproductive units are all the same in size and structure, unlike the large eggs and small sperm of species that have sexual differentiation into male and female forms. The isogametes swim about in the seawater and fuse together in pairs, producing cells known as zygotes, from which the organisms develop. Like fertilized *Fucus* eggs, these zygotes secrete a sticky substance and attach to rocks. The first sign of the emergence of structure is the formation of a little tip that grows up toward the light and irregular outgrowths from the base that form the rootlike rhizoid, in one of whose branches the nucleus takes up residence. The tip grows and grows, and what was an invisible cell, only a few thousandths of a centimeter in diameter, begins to take on visible dimensions of several millimeters. Then, at the tip, new structures suddenly arise. These are rings of little leaflike elements called *bracts* or *laterals,* each of which grows and branches, the whole structure being called a *verticil* or a *whorl,* as can be seen in Figures

79

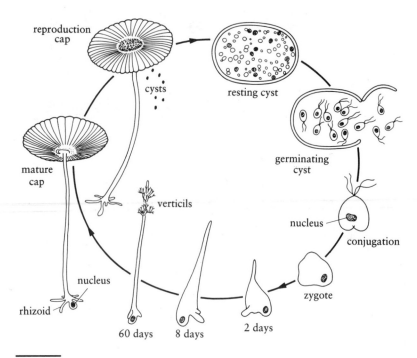

Figure 4.2   *The main stages of the life cycle of* Acetabularia acetabulum, *from zygote formation through stalk growth; verticil or whorl formation to the emergence and growth of the cap to give the mature form; and cyst formation and release of isogametes as the cycle starts again.*

4.2 and 4.3. From the center of a whorl the tip starts growing again, and, after several days during which the tip grows a few millimeters, another whorl of laterals is produced. This process is repeated many times, the cell becoming festooned along its length with the delicate tracery of a sequence of whorls that then drop off in sequence, oldest first, rather like the shedding of leaves from a tree. Finally a different structure is produced, a little crenulated knob whose elements remain connected instead of separating like the laterals of a whorl. This structure grows in diameter, and the beautiful sculptured detail of the cap begins to emerge, developing into the distinctive form that characterizes the species. This whole process, in which the simple little zygote transforms through a well-defined sequence of stages into the adult form, is called *morphogenesis* (creation of form). The adult form of

Figure 4.3   *A developing alga with whorls, showing how the individual laterals branch. The tip continues its growth from the center of a whorl and produces a new whorl every few days. They then drop off.*

the alga consists of a rhizoid with the nucleus in one of its branches, a thin (approximately 0.5 millimeter) stalk some 3 to 5 centimeters in length, and a cap about 0.5 centimeters in diameter (Figure 4.4). That is a pretty impressive size for a single cell! Other members of the group are even bigger, and in a moment we will look at some of *Acetabularia*'s relatives to get an idea of the diversity of forms they can have. But first let's complete the life cycle.

The adult cell lives for several months, usually in clumps attached at their roots as shown in Figure 4.1. The cell surfaces—the walls— tend to develop a whitish tinge as they age due to the deposition of calcium from the seawater. There comes a time in the age of an *Ace-tabularia* cell, between three and six months, when it decides to go into reproductive mode. The single nucleus divides many times, pro-ducing thousands of nuclei that travel up into the cap. There the nuclei, which have become haploid (half the chromosome number of the zygote), produce little flagellated gametes that then get packaged into cysts, containers constructed with trapdoors that later open like the

Figure 4.4   *The adult form of* Acetabularia, *showing the structure of the cap, the stalk, and the rhizoid.*

hatch of a deep-sea diving bell. Because the gametes are produced within the cap, the latter is known as a gametophore. The cell wall then dissolves away, releasing the cysts. These drift about, the trapdoors open, and the isogametes swim out into the seawater. Many algae mature simultaneously, so dense clouds of the tiny green swimmers are produced, looking like a typical algal bloom. They fuse together in pairs, making zygotes, and the life cycle begins again.

Every species has a variant of such a cycle, some being extremely complex, such as those of parasites that live different stages of the cycle in different hosts so that they have to find ways of getting from one to the other. Compared to these, *Acetabularia* has a very simple reproductive cycle, its claim to fame being the growth of a single cell to such an extraordinary size and delicate beauty. But it has relatives that exceed it in both respects. *Acetabularia major,* living in the Pacific Ocean around the Hawaiian islands, grows up to 10 centimeters long. Members of the genus *Batophora* (Figure 4.5) have even more complex

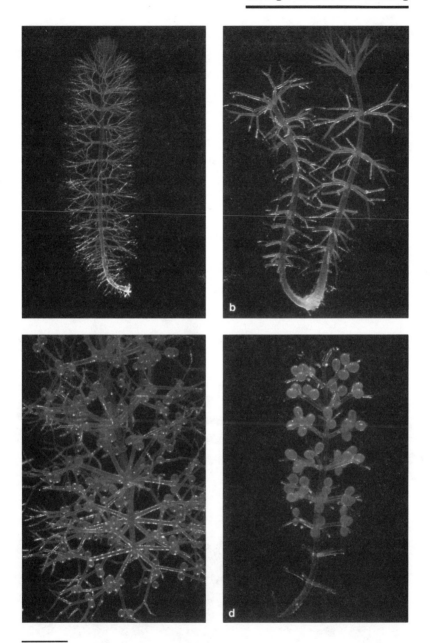

Figure 4.5   *Two species of the genus* Batophora: *whorls of laterals are retained in the adult (above); gametophores are produced on the laterals (below).*

patterns of whorls, which remain on the adult. These species do not make caps, the gametes being produced instead on the laterals in little ovoid structures. In other species, the gametes are produced within the laterals themselves, as in *Halicoryne spicata* (Figure 4.6). The whorls clearly have a function in these species. But this raises the question of why *Acetabularia acetabulum* spends so much energy producing whorls of laterals that simply drop off, serving no function in the adult where the cap takes over as the main site of photosynthesis and as the gametophore. In looking for an answer to this question we are going to find a type of explanation rather different from the usual ones based on inheritance, usefulness, and function that contribute to the selfish-gene view of organisms and their survival. To start on the path that will take us in a new direction, we can begin with the innocent little question: Why does *Acetabularia* make whorls?

Figure 4.6   *The species* Halicoryne spicata *with whorls that themselves become the gametophores.*

History and Structure

One of the traditional approaches to "Why" questions in biology is to take a historical perspective. This, after all, was Darwin's vision: species arise from one another by the inheritance of small changes, so the properties of an ancestor of a group are expected to leave their trace in the descendants. This is an argument based on historical continuity and a kind of inertia, a resistance to change. Did the ancestors of *Acetabularia* have whorls? Indeed they did. The fossil *Dasycladales* all produced laterals, and in the great majority they were arranged in whorls. So this is an ancestral character. The proposal is that *Acetabularia* gets its whorls as a historical legacy. But if it has no use for these structures, shedding them soon after they are made, why does it not just get rid of them altogether? The answer is developmental inertia: for some reason, making whorls is just too deep and persistent a property of morphogenesis to get rid of, and the "cost" of making these structures is not a sufficient penalty to reduce the survival capacity, or fitness, of this and all the other species that make whorls but do not use them. These arguments sound perfectly reasonable and plausible, but they point to a problem rather than to a solution. There is something very strange about accepting them as a satisfactory scientific explanation of the phenomenon. And yet this type of argument is widely accepted in biology. Let's look at an example from physics to put the biological problem in a different perspective.

Imagine a time when you didn't know why Earth goes around the sun in an elliptical orbit. You ask someone to give you an explanation, and they say: "Earth goes around the sun in an elliptical orbit this year because that is what it did last year, and the year before that, and so on back to the origin of the planetary system, and nothing has happened to change it." Are you satisfied? The answer is perfectly correct; it is a historical explanation of the type that is widely used in biology. Trace something back to its origins, point to the first known instance of the phenomenon and use that as the explanation of phenomena subsequently connected to this event. In biology this origin is usually identified as the common ancestor of some lineage. It comes from the study of genealogies and family trees, whose origins are identified with significant individuals such as an ancestor who first came to America from England, say, or William the Conqueror as the first Norman king of Britain. But it can be applied to anything else, as long as there is an adequate historical record to trace back the sequence of events to the origin. Hence the importance of the fossil record in biology.

Now let's return to Earth's orbit. Do you accept the historical explanation as adequate? It is both reasonable and correct. But ever since Newton, this type of explanation has been regarded as inadequate. The reason is that it gives no insight into why the form of Earth's orbit is elliptical. All that we are told is that it started that way, and that's how it has continued, through a kind of inertia, with nothing happening to change it significantly. What Newton did was explain why it is an ellipse. He did so by calculating what shapes of orbit a planet such as Earth can have if it is rotating about a massive body like the sun, assuming that the force acting between them obeys the inverse square law of gravitational attraction. He discovered that the only possible forms of planetary motion are the conic sections: circle, ellipse, parabola, hyperbola. Of these, only circles and ellipses have repeating or periodic orbits (circles being in fact just special cases of ellipses). Parabolas and hyperbolas involve motion that goes off to infinity (in a Newtonian universe, though of course orbits are different in an Einsteinian world with curved space). Kepler had shown that Earth's orbit is an ellipse, and we still need to explain why Earth follows this orbit rather than a circle. This is where history comes in, as it does into all dynamical problems. The gravitating mass of gas and dust out of which the sun and the planets developed was rotating at a velocity less than the escape velocity needed to send the planets on a space trip. So they went into closed orbits around the sun. To get circles, the initial velocity has to be just right—they are possible, but improbable. Ellipses, on the other hand, are much more probable, so they are the expected outcome of planetary evolution according to Newtonian dynamics. History comes into this explanation in the form of initial conditions. Every dynamical problem has to start somewhere, and these initial conditions are the bit of history that enters the problem. In the case of planetary evolution, it is the mass and rotation velocity of the cloud of matter from which our solar system developed, the initial velocities and masses of the planets, that determine their exact orbits. However, the goal of scientific explanation is to reduce accidentals to the minimum possible and to seek the principles according to which systems are organized. The procedure is to describe the states that are possible in a system according to its intrinsic properties and then to explain what is observed as a specific actualization of the possible by the action of particular conditions ("history"). Historical explanations are not regarded as sufficient. Why, then, are they so extensively used in biology?

The basic reason goes back to an assumption made by Darwin that is still very widely accepted. As we saw in chapter 2, Darwin believed that organisms undergo small, random hereditary changes, among which is selection of the better adapted variants. So evolutionary change depends on continuous variation. Now here comes the additional assumption: these small variations are such that almost *anything* can happen—organisms can take any form, have any color, and eat any food, subject only to very broad constraints that are basically due to physical and chemical laws. Elephants will never be able to fly because there are no materials that would allow them to develop wings sufficiently light and strong, and muscles sufficiently powerful. And organisms cannot eat diamond despite the fact that it is made of carbon, though they do "eat" iron in the sense of using it in their energy-generating activities, as do certain types of bacteria.

If it is true that organisms have a dense spectrum of states that they can occupy, almost anything being possible for evolution by natural selection, then organisms do not obey rules in the way that physical and chemical systems do. Physical laws result in only certain possible forms for planetary orbits and for liquids (as described in chapter 1). However, if organisms can do almost anything, then they do not obey any laws of their own; that is, there are no principles of organization in biology that might allow us to describe what is possible and what is not in the biological realm. Then the whole of biology is really just the accidents of history: what forms happened to be selected in particular environments. Biology thus becomes a set of historical narratives: which species come from which ancestors under which circumstances, with the only necessity being survival. Unlike the other sciences, in which principles of organization allow one to understand the structure of the physical and chemical world in terms of regularities and general principles, the phenomena of biology are *unintelligible* in such terms, and survival is the only law. This is why natural selection has become so important in biology: it is the only "force" that is used to explain what has happened during evolution.

The trouble is that natural selection provides a very limited type of explanation, and it fails completely on some very important and interesting questions. Going back to the case of whorls, all that natural selection can offer by way of explanation is that whorls are useful in most members of the giant unicellular green algal order as gametophores, and they evidently do not "cost" too much in *Acetabularia,* so they keep on being produced even when they are not used to make gametes. Of course, whorls also carry out photosynthesis during the growth of the cell, and they may serve other functions that we are not aware of—one can never be sure. But during most of the life cycle of *Acetabularia* they are absent. Explanations in terms of history and natural selection are not very helpful since they merely redescribe what is observed in terms of functions and costs, but one is no wiser for the "explanation." Voltaire was scathing about doctors in the eighteenth century who described the efficacy of sleeping drafts such as laudanum in terms of their "dormative principles." So no one should be under any illusions about the value of biological explanations in terms of historical narratives and natural selection. But what is needed to go beyond these?

If we are going to go deeper than history and function to an understanding of form or structure, then it is clear that we must attempt to develop a theory of morphogenesis: how organisms with their distinctive forms are produced. For that is what is involved in producing an explanation of the type provided by Newton for planetary motion, which at the same time accounted for the fall of the apple, the tidal rhythms of the oceans, and countless other phenomena, all with the same basic principles. What we are about to embark on is a similar adventure, looking for the general principles of biological organization that could account for the forms of life. This sounds ridiculously ambitious, and it is. But in science one does not usually start with the big question; one starts with a small, puzzling one. And one finds out that there have been, and are, a lot of other people who have contributed important parts to the puzzle. When these begin to coalesce into a general picture, excitement starts to rise. A number of people currently working on various aspects of the sciences of complexity are beginning to feel this coalescence in a subject that is still quite amorphous and difficult to describe precisely (I say a lot more about this in chapter 6). At the moment, it is useful to look at a little problem where the issues are sharp and precise. Once again: why does *Acetabularia* make whorls?

The Parts of the Puzzle

One of the attractive features of *Acetabularia,* apart from its delicate beauty, is that it is a simple organism that is nevertheless sufficiently complex to present us with a nontrivial problem in morphogenesis. This organism has only four major parts: the wall, the cytoplasm, the nucleus, and the large fluid-filled chamber called the *vacuole* that occupies most of the interior of the cell. The cytoplasm is where most of the activities of the cell occur. It is made up of many different components, including chloroplasts and mitochondria (the energy generators). In *Acetabularia* it forms a thin shell between the inner vacuole and the wall, which is an even thinner shell on the outside. A considerable concentration of salts and small organic molecules in the vacuole results in a strong osmotic pressure pushing outward on the wall, so that the cell maintains its shape in the same way as an inflated balloon. If this pressure drops, the cell wilts, just like a plant that needs watering, and for exactly the same reason, for it is water pressure that maintains turgor. As far as the production of shape is concerned, this is all we need to know about the role of the vacuole: it exerts pressure on the wall.

What about the wall itself? Clearly it grows and changes its shape as the cell progresses through different stages of its life cycle (Figure 4.2). Remember that the whorls and the cap are not made of separate cells, but result from intricate changes of shape in the wall itself, which forms a continuous shell around the whole organism. Is it the wall itself that initiates these shape changes and organizes the morphogenetic sequence that gives rise to the adult form of the organism? The wall is a relatively inert structure that responds to influences located elsewhere that determine where the wall grows and how it changes its shape, so it does not appear to be the source of the spatial patterning. There are only two candidates left: the thin layer of cytoplasm that lies between the wall and the vacuole, and the nucleus located within the cytoplasm in one of the branches of the rhizoid. Which of these organizes the development of spatial patterns in the growing alga? There is no doubt what the majority opinion would be on this: the nucleus has the information, or plan, for generating the shape of the organism. Well, there is a nice experiment that can be done to test this hypothesis.

If an adult *Acetabularia* is cut in two by simply snipping through the middle of the stalk, the lower (basal) part with the rhizoid and the nucleus regenerates a new cap, while the part with the cap lives for several weeks and then dies. Regeneration from the basal part takes place by the same sequence of events that characterizes normal morphogenesis: after the cut has healed with a new cell wall formed over the cut, a little tip forms, grows, and produces a series of whorls and then a cap. The whorls drop off and the alga is as good as new, indistinguishable from the original. This organism can regenerate its form, and is good for studying morphogenesis because the same plants can be used again and again. They don't wear out! This experiment points toward the nucleus as the organizer of morphogenesis, since it is only the part of the alga with the nucleus that regenerates. However, there is another experiment that can be done, whose results make us think again.

Suppose that the cap is cut off, along with the rhizoid, leaving only the stalk. We know what happens to the part with the cap—it lives for a while and then dies. And the basal part with the nucleus regenerates the whole alga. What about the stalk? The expectation is certainly that it will behave like the cap—live for a while, then die. And that is correct. But what it does while it is alive is unexpected. It regenerates a cap! And it does so in the usual way: healing at both ends, growing a tip where the cap used to be, which produces whorls and then a new cap. Occasionally, regeneration occurs at the opposite end of the stalk, where the base used to be, and very occasionally from both ends, the result being a stalk with a cap at both ends. However, a part without a nucleus always dies eventually. What are we to make of these intriguing observations?

First, a nucleus is clearly necessary for the survival of the organism and for the completion of the life cycle. It is equally clear that morphogenesis occurs in the absence of a nucleus, so it is not the nucleus itself that directs this process. However, it is easy to show that nuclear *products* are required in the cytoplasm for regeneration to take place. If the cap that is regenerated by a stalk without a nucleus is cut off, so the stalk loses its cap for the second time, the stalk is unable to grow another one, getting no further than a tip and a whorl or two. Evidently, the nuclear materials get used up during regeneration, a conclusion that can be confirmed by direct measurement of nuclear materials in the cytoplasm. These products are made in the nucleus, released, and distributed to all parts of the cell by active streaming movements of the cytoplasm. In such a big cell, diffusion alone would be inadequate to get these large informational molecules (messenger RNA) and the proteins they produce to the places where they are needed for the construction of structures such as laterals and a cap. But what organizes the actual construction? This seems to be the job of the cytoplasm.

We now need more information about the properties of the cytoplasm that could indicate how the various changes of shape in the developing organism might occur. One property is suggested by the effect of calcium on morphogenesis. Work in my laboratory has shown that if the concentration of this ion (the charged form of calcium, Ca^{++}) is changed, the pattern of morphogenesis can be altered. In seawater, calcium is normally at a concentration of about 10 millimoles (which means one-thousandth of a mole per liter). If this is reduced to 1 millimole while all the other components of seawater are kept at their normal concentrations, then the cells are unable to regenerate after their caps are cut off. At 1.5 millimoles of calcium, tips form and the stalk grows, but no whorls are produced. Under these conditions the cells just go on growing indefinitely, failing to complete normal morphogenesis. At 2.5 millimoles of calcium, the stalk grows and a sequence of whorls is produced, but no caps are formed. Not until the calcium concentration reaches 4 millimoles or greater do caps begin to be produced. Various experiments showed that what was important was the amount of calcium getting into the cytoplasm, not just its effect on the cell wall. Detailed observations by Lionel Harrison and his colleagues in Canada showed that small changes in calcium concentration affected the spacing between the laterals in a whorl. So it is clear that this ion is important in the morphogenesis of *Acetabularia*. Lionel Jaffe, working in the United States with *Fucus* eggs, has shown that the first sign of the axis of the alga is an electrical current produced by an inflow of calcium into the part of the cell where the outgrowth occurs. This ion has turned out to be involved wherever morphogenesis, change of shape, is occurring, irrespective of the organism—plant or animal. So the effects of calcium on the cytoplasm seem to be important in understanding how cells change their shapes.

The cytoplasm is not just a blob of soft jelly with a lot of things dissolved in it. It actually has within it a complex and intricate structure in the form of a network of protein polymers that make up the cytoskeleton (*cyto* means cell; see Figure 4.7). These polymers are constantly being made and broken down again, so the whole structure is extremely dynamic. It is, in fact, quite chaotic in its activities. The

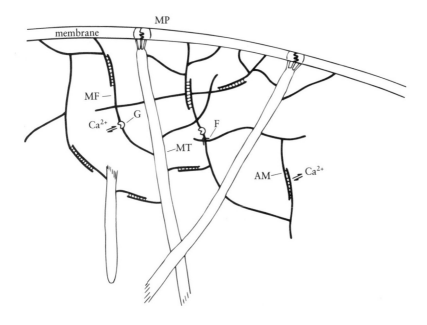

Figure 4.7   *Some elements of the cytoskeleton: microtubules, microfilaments, actomyosin, and various proteins that act on the
cytoskeleton and anchor it to the membrane, together with
some of the sites of calcium action.*

mechanical state of the cytoplasm—its rigidity or softness—depends
largely on the state of the cytoskeletal polymers. And one of the main
factors involved in making and unmaking the polymers is calcium. So
calcium influences the mechanical properties of a cell; a mechanical
property of the cytoplasm called the *elastic modulus* (a measure of
stiffness or resistance to deformation) is a function of the calcium
concentration. Figure 4.8 shows this relationship. The effect of different concentrations of calcium on *Acetabularia* morphogenesis described earlier is due to these effects on the cytoskeleton as well as the
influences of calcium on the elastic modulus of the cell wall. However,
we need a more precise description of how these influences act. For
this we need equations that describe, on one hand, the dynamics of
calcium variation in a cell, and on the other the mechanical properties
of the cytoplasm, together with their interactions.

93

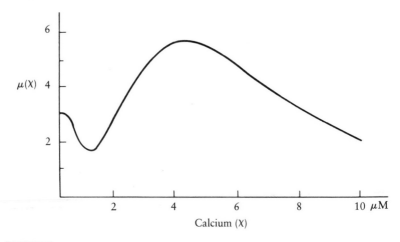

Figure 4.8 *Elastic modulus of the cytoplasm varies with the concentra-*
tion of free calcium. As calcium increases from its normal
concentration of 10^{-7} moles, the cytoplasm first softens (the
elastic modulus decreases) as enzymes that cut microfilaments
are activated, and then it stiffens as actomyosin contracts
(elastic modulus increases). Finally, above 5×10^{-6} moles of
calcium, there is again softening as microtubules depoly-
merize and microfilaments are cut into ever shorter units.

Morphogenetic Fields

I am now going to talk about morphogenetic fields that have a definition as precise as that for any field used in physics, so it is important
not to confuse them with the concept of morphic resonance used by
Rupert Sheldrake in his book *A New Science of Life.* His fields are
nonphysical, whereas the concept of field used to describe pattern
formation in biology for many decades refers to spatial organizing
activities that involve clearly defined physical and chemical processes,
though combined in a way that is distinctive to the living state. The
derivation of the equations that describe the morphogenetic field I now
discuss was carried out in collaboration with Lynn Trainor, a physicist
at the University of Toronto, and Christian Brière, a biologist and
computer scientist who works at an agricultural research institute in
Toulouse, France. We needed equations that describe how a cell controls the concentration of calcium in the cytoplasm, how forces acting

94

on the cytoplasm affect its state, how this state influences calcium concentration, and how calcium affects the mechanical properties of the cytoplasm through its influence on the cytoskeleton. Cells exert very stringent control over the concentration of free calcium, because this ion has strong effects on a large number of cell functions due to its capacity to bind to proteins and modify their activities. Calcium is, in effect, a potent poison, and if it is allowed to increase much above $0.00001 = 10^{-5}$ moles, all the vital functions of a cell are so seriously disrupted that the cell dies. The concentration is therefore controlled within the range 10^{-7} to 10^{-5} moles, usually toward the lower end of this range. The ion is bound by special proteins that hold it in a nonreactive state, and it is pumped into storage chambers within the cell or out across the cell membrane. This very precise control over free calcium within cells means that this ion becomes an exquisite control signal, affecting mechanical properties through its effects on cytoskeletal proteins, and metabolic activities via its influence on enzymes, as shown in Figures 4.7 and 4.8.

The equations that Trainor, Brière, and I derived for these interactions depended on the work of many others who had developed mathematical descriptions of morphogenesis, as well as the experimental results of investigators on the properties of the cytoskeleton and calcium, and on the growth of plant cell walls. In particular, we followed an interesting trail blazed by Gary Odell at the Rensselaer Polytechnic Institute in Troy, New York, and George Oster, of the University of California at Berkeley. They showed how calcium–cytoskeleton interactions could result in the spontaneous formation of spatial patterns in the concentration of free calcium and the mechanical state of the cytoplasm, measured by strain (stretching or compression). Our assumptions for *Acetabularia* were a bit different from theirs, but we found the same property: starting from spatially uniform states, patterns of calcium and mechanical state emerged spontaneously as a result of the interactions between them. This type of behavior is absolutely crucial for any model of morphogenesis, because, as described in the preceding chapter, a developing organism has to generate its own form from a simple, symmetrical initial shape. In *Acetabularia* the spherical zygote has to break out of its simplicity into ordered complexity of form. The technical term to describe the transition from a state of higher symmetry (lower complexity) to one of lower symmetry (higher complexity) is *bifurcation*. Why does the dynamics of the calcium–cytoskeleton interaction have this property? You won't be surprised to learn that it is the same reason as for the emergence of pattern in the Beloussov-Zhabotinsky reaction, aggregating slime mold amoebas, and all the other systems described in the preceding chapter. There is a positive feedback process that increases the concentration of calcium, arising from its effects on the cytoskeleton, and there are opposing processes that decrease it. Calcium is not synthesized in a chemical reaction the way bromous acid or cAMP is, since it is an element, and only stars can make it. However, it can be released from a bound or stored state into the cytoplasm where it affects the cytoskeleton by the sorts of interactions shown in Figure 4.7. The cytoplasm can be regarded as an elastic skin around the vacuole, kept under tension (stretched) by the osmotic pressure from the vacuole pushing on the cell wall, against which the cytoplasm is closely apposed.

Wherever free calcium increases slightly from its resting state of 10^{-7} moles, the cytoskeletal polymers are affected and the cytoplasm softens (see the initial decrease in the elastic modulus of the cytoplasm in Figure 4.8). Therefore, it stretches so that the *strain* (the technical term for the amount of stretch in a material) increases. This increased strain causes release of calcium from its bound or stored form, a result observed experimentally and built into the model. So more free calcium accumulates, the cytoplasm softens more, and more calcium is released. This positive feedback loop is potentially explosive, as are all such loops, and if the process were to continue, the cytoplasm would stretch to the point where it tore. However, this process is opposed by others, such as the diffusion of calcium away from regions of accumulation, its removal by pumps, and the stiffening of the cytoplasm as calcium concentration rises beyond about 10^{-6} moles (the increase of the elastic modulus after its initial decrease in Figure 4.8). We have here the combination of positive and negative feedback processes that give the cytoplasm the ingredients of an excitable medium. So what we could expect to find is a variety of dynamical states that the cytoplasm has available to it: (1) a uniform steady state in which everything is in balance and no patterns emerge; (2) spontaneous bifurcation, or symmetry breaking, from an initially uniform state to a stationary wave pattern of calcium and strain with a characteristic wavelength; (3) a similar wave pattern but now dynamic, with peaks and troughs of calcium and strain propagating across the cytoplasm in either concentric circles or spirals; and (4) chaotic patterns in time and space. These are the main patterns generated by excitable media and we encountered all of them in chapter 3 except for the second one listed, which is going to be one of the most important for morphogenesis. They all appeared in our computer simulations, each one's presence dependent on the values of the constants (the parameters) in the equations. These are quantities such as the affinity of the binding proteins for calcium, the effective diffusion constant for calcium, the resistance of the cytoskeletal elements to deformation (i.e., the restoring force), and so on. All of these are quantities that are affected by gene products, so they can change from one species to another, and indeed they can change during the development of an individual if gene activity changes in ways that affect these parameters. As parameters change, different patterns arise. This is one way in which genes can affect morphogenesis. The environment also influences development via such quantities as the concentration of calcium in the seawater, which is also taken as a parameter in the equations.

Computer simulations allowed us to explore the types of form that a cell like a developing alga could produce. We had here a model of a morphogenetic field that could tell us what shapes are typical of (generic to) this kind of organism. Would we find whorls emerging as a natural form? At this stage of our investigations, I confess that I considered this to be a rather remote possibility. While our model was biologically very simple—vacuole, cytoplasm, and cell wall—it was nevertheless very complex mathematically, with no fewer than twenty-six parameters. From the study of simpler systems, and particularly from linear approximations to their behavior in which the properties of the whole are, roughly speaking, the sum of the properties of the parts, a saying had emerged: If you have more than three parameters in your model, anything is possible! This suggested that we were in a Darwinian world where anything could happen, any shape could emerge, and it was simply a case of selecting the right combination of parameter values (genes) to get any form we wanted. Conventional wisdom had it that evolution had plenty of time to explore this immense space of possibilities by random variation of parameters and discover structures that worked in the prevailing circumstances (the habitat). The early *Dasycladales* hit upon whorls of laterals because they worked as leaflike structures that combined photosynthetic activity with the reproductive function of gametophores. *Acetabularia* never bothered to get rid of them when it discovered how to make caps, except by making and then discarding them. We might have been in for a very slow, tedious search of an enormous parameter space in an attempt to repeat what evolution had achieved over millennia— except that we did not have such time spans at our disposal.

Generic Forms

What we were looking for was a sequence of shape changes like that shown in Figure 4.9 (the forms produced after the cap is removed, up to the first whorl of laterals). The cut heals; a new cell wall forms a hemisphere over the cytoplasm, pushed out by vacuolar pressure; a tip arises, a stalk grows out of it, then the tip flattens and the beginnings of a whorl emerge. Our model included interactions between the state of the cytoplasm and the wall that depended on the strain in the cytoplasm: the wall softened wherever the cytoplasm was strained beyond a critical value, representing a process that had been identified and characterized experimentally in plant cells. In addition, the model included equations describing the growth of the wall wherever its strain

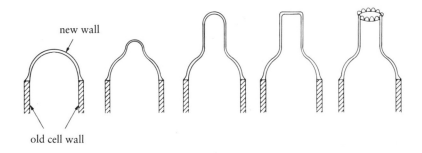

new wall

old cell wall

Figure 4.9   *Shape changes that a developing or a regenerating alga undergoes, leading to the production of a whorl.*

exceeded a critical value, so that not only did the elastic modulus (the stiffness) of the wall change in response to cytoplasmic state, resulting in elastic deformation, but the wall could also undergo plastic changes through differential growth, which is of course what plant cell walls do. We set the parameters to values within the bifurcation range that allowed the spontaneous emergence of spatial patterns in calcium concentration and cytoplasmic strain, with wavelengths that were smaller than the size of the regenerating hemisphere, so that structures could develop within this domain. Then we let it do its own thing.

The first pattern to emerge was a gradient of calcium and strain in the cytoplasm, with maxima at the tip (Figure 4.10). This computer graphic shows increasing calcium concentration toward the tip. The three-dimensional image is produced by finite element analysis: each of the little lines in the figure represents the cytoplasm and obeys the equations of the model. There is another shell of similar shape that represents the cell wall.

When the strain got sufficiently large, the wall at the tip softened, and we had the first sign of a changing structure in the regenerating hemisphere (Figure 4.11). This was followed by growth of the tip and stalk formation. These were the sorts of processes that we had hoped for and expected, because there are limited ways in which the spherical symmetry of the regenerating hemisphere can be broken in the initi-

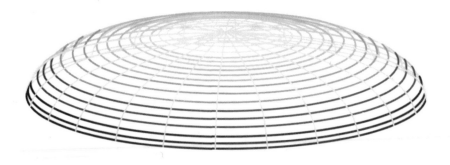

Figure 4.10 *Gradient in free calcium concentration that is produced spontaneously in a regenerating alga, with a maximum at the apex of the dome. The lines that make up the structure are the finite elements that are used in the analysis.*

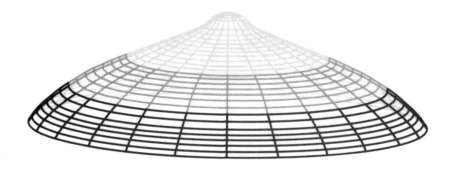

Figure 4.11 *Tip formation in the model, showing the elastic deformation that initiates growth of the tip.*

ation of morphogenesis, and this was one of them. Still, it was reassuring to see that the model was reproducing so faithfully the initial stages of the regeneration process.

Then came our first real surprise. After a certain amount of growth, the tip suddenly began to flatten. This was something that we had often seen in growing and regenerating algae, and it was always a sign that a whorl was about to be produced; but we had never understood why it occurred. Now the model suggested an answer. When we looked at the calcium pattern, we saw that it had changed from a gradient with a maximum at the tip, as in Figures 4.10 and 4.11, to the pattern shown in Figure 4.12: now the maximum is back from the tip and the calcium profile is in the form of a ring, or an annulus, rising to a peak and then falling again at the tip. Cytoplasmic strain shows a similar pattern. The result is that the wall gets softened on the annulus so that this is where it curves most in response to pressure from the vacuole, while the tip gets stiffer and flattens. So here was an expla-

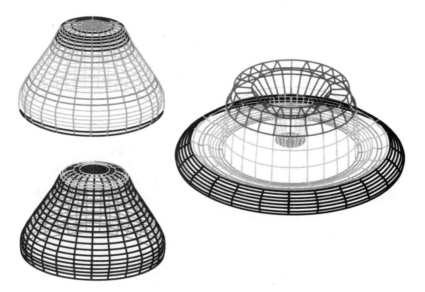

Figure 4.12   *Growth in the model, resulting in the formation of an annulus of calcium and strain, with maxima as shown (right). This pattern of calcium and strain results in tip flattening (left).*

nation of this characteristic change of shape. But the next question was the big one: would a whorl now form on this annulus?

This is actually a stability question: is the annulus stable to random perturbations, or is there a nearby state that it prefers to settle into? We tested the stability of the annulus by giving it a shake, that is, introducing perturbations into the calcium concentration so that it no longer had a uniform value around the annulus. It changed into the pattern shown in Figure 4.13. This has the symmetry of a whorl: a series of peaks around a flattened tip. But it is a pattern of calcium, not an actual whorl, which consists of a crown of little growing tips each of which then develops into a branching lateral. Our model could do no more than this because of a technical limitation. Each of the little peaks in calcium, which is accompanied by a similar peak in strain, should behave just like a tip, but on a smaller scale: the wall softens over the peak, bulges out under pressure, and grows into a lateral. But this requires a shift of scale in the model, which is constructed out of finite elements—the little lines that make up the overall

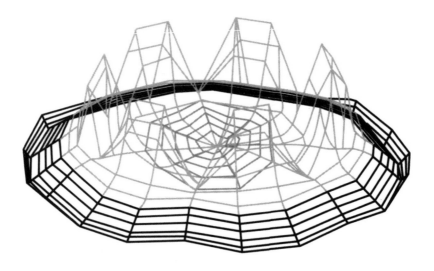

Figure 4.13   *Under perturbation, the calcium annulus transforms into the ring of peaks shown, which has the symmetries of a whorl and initiates whorl formation.*

structure of the regenerating tip. The model has two of these shells of finite elements, one for the cytoplasm and one for the wall. Each of these elements is dynamic—it obeys the equations of the model and stretches, shrinks, or grows according to the values of the variables that define its state. The computer program calculates each of these in succession, for both shells—a lot of computing! To grow a whorl of laterals, the whole program would have to be repeated on a finer scale, and many times over. We did not have either a computer or the technical personnel to carry this out. So we had to be content with the first stage of whorl formation in the model. However, it was enough to show us that the morphogenetic sequence we were trying to discover was extraordinarily easy to find. We did not have to spend the equivalent of millions of years of evolution before falling upon it. Perhaps evolution did not have to look very hard either. There seems to be a large domain in the space of parameters that leads to whorls; that is, as long as parameters are set anywhere within a large range of values the system will converge on the whorl structure. However, to draw that conclusion with certainty requires a much more extensive study than we have been able to carry out, or a shift of tactics to a more mathematical approach that can give us a better feel for what is going on dynamically in the model. Nevertheless, the ease with which we found the whorl pattern in our model certainly points in the direction of a large region for this structure in the parameter space representing morphogenesis in unicellular algae. That implies that this structure is actually highly probable. Genes do not have to be very precisely controlled to fall within the region where whorls are produced. For a system with the basic organization of *Acetabularia,* which all members of the *Dasycladales* have in common, whorls of laterals appear to be generic forms—structures that are typical for this type of morphogenetic process, as ellipses are for planetary motion in gravitational fields. I have to stress that this is *not* a proven result; it is a conjecture. However, the plausibility of this conjecture will increase as different lines of evidence are pursued. For the moment, let's return to the model, for further results suggest that we were on an interesting track.

Although we could not actually grow whorls, we could let the model go on growing to see whether there was a repetition of the whorl pattern during stalk growth, as occurs in the algae with production of a series of whorls. We found that there was indeed a repeat of the sequence: calcium gradient formation with maximum at the tip and growth of the stalk; then transformation of the calcium gradient into an annulus with a peak a little way back from the tip, resulting in characteristic tip flattening and the potential for whorl formation by the emergence of a ring of peaks out of the calcium annulus, as in Figure 4.12. This process is like a traveling wave that rises and falls with an irregular periodicity, leaving a structural record of its passage in the sequence of whorls. What we are dealing with here is what is called a *moving boundary problem,* described by a category of field equations in which the boundary of the field moves as a result of patterned growth, as in a growing crystal or, in our case, a growing cell. This type of process is of particular importance in morphogenesis, since the organism changes its shape as it develops. The result is that there is an intimate connection between the dynamics and the form: the field dynamic generates a pattern that leads to a particular shape that then affects the dynamics, resulting in an unfolding of form through a sequence of changes. We can describe this as an implicate order in the dynamics being explicated in the shape, which then influences the implicate order. Although in our model we kept parameters constant, it is known that gene products such as enzymes do change during morphogenesis in *Acetabularia,* and these can alter parameters. So in the real organism the dynamics is more complex than in our model. However, because we do not know exactly what genes are involved and how they act through the physics and chemistry of morphogenesis, we preferred to work with a simpler model and see how far it could take us.

We failed to find conditions that gave a structure like the cap of *Acetabularia.* What we did get was a large, bulbous terminal structure, not the parasol-like form of the Mermaid's Cap. Evidently this form is not so easy to find, and it is quite likely that changes will be required in our model before such a shape is generated. Remember, however, that caps are latecomers on the scene of *Dasycladales* evolution. Whereas all members of the order produce whorls of laterals, only the relatively recent group known as the *Acetabulariaceae* produce caps. So it is likely that caps are produced only when parameter values (genes) are in a fairly restricted range, representing a low-probability structure in algal morphospace, the space of potential forms.

What about the experimental evidence? Is it consistent with the model? Lionel Harrison and his colleagues at the University of British Columbia have reported that calcium is maximal at the tip of a growing *Acetabularia* cell, that it then forms into an annulus when the tip flattens, and that the pattern transforms into a series of peaks corresponding to the individual laterals of a whorl. They were looking at calcium bound to membranes, not the free form of the ion, but it is to be expected that these will vary in a similar manner. These observations are consistent with our model and with other morphogenetic models, one of which has been proposed by Harrison himself. Instead of studying calcium interacting with the cytoskeleton, he assumed that interactions between chemical reactions were the primary generators of dynamic patterns, similar to those used to describe the dynamics of signaling in the slime mold. These include reactions involving calcium, which plays a significant role in the aggregation process, as it does also in muscle and nerve activity. As mentioned earlier, calcium acts in so many basic cellular activities that it is bound to be one of the variables in any of these excitable systems, and it may be difficult to decide exactly what role it plays, whether primary, as in our model, or secondary, as in Harrison's. Despite differences of emphasis, all morphogenetic field models are, in a fundamental sense, equivalent. They depend on excitable dynamics to initiate large-scale or global pattern formation.

The first person to demonstrate that chemical reactions, combined with diffusion, could produce spatial patterns by spontaneous symmetry breaking of an initial spatially uniform state was the mathematical genius Alan Turing, better known for his brilliant work on the logic of computation and the development of the first powerful computer the world ever saw, in Manchester just after World War II. He was intrigued by biological patterns such as leaf arrangements and the spots on butterfly wings and Dalmatian dogs, and proceeded to show how biochemical reactions of sufficient complexity (nonlinearity), together with diffusion of their products, could produce stationary wave patterns of the chemicals involved in the reactions. His paper "The Chemical Basis of Morphogenesis," published in 1952, was the beginning of the study of excitable media, although this term did not come into prominence until much later, with the discovery of the full range of dynamic patterns that these systems can produce. Ilya Prigogine and his colleagues in Brussels have made important contributions to this area of study, and the concept of "order from chaos" in complex systems is a prominent theme in their work (see *Exploring Complexity*). There are now many different types of systems, both animate and inanimate, that have been shown to have the property of bifurcation from spatial uniformity to pattern. They all share certain properties such as nonlinearity, energy flow through the system so that they are displaced from thermodynamic equilibrium, and chaotic patterns of fluctuation at bifurcation and transition points that are diagnostic of these sudden changes of state. We considered a few of these in the preceding chapter, and there are many more. A recent treatment of some fascinating examples from a perspective that is particularly germane to the theme of this book is *Fearful Symmetry: Is God a Geometer?* by Ian Stewart and Martin Golubitsky. These authors examine symmetry-breaking sequences that generate patterns of complex dynamic order in living and nonliving systems, including fluid flow patterns, the forms of galaxies, and Turing patterns in morphogenesis.

Robust Morphogenesis

We now need to take a further step in identifying a property that is crucial to understanding the distinctive qualities of developing organisms as complex systems of a particular kind, and the extraordinary robustness of the forms of life. This property seems to provide a clue to one of the most fundamental and striking aspects of the living realm, and lies at the foundation of all biological study: the fact that different types of organisms can be related to one another in systematic, orderly patterns that are called *taxonomies*. The father of biology as we know it today was the great Swedish taxonomist Carolus Linnaeus. Organisms are not random assemblages of working parts, the results of trial-and-error tinkering by natural selection. They reflect a deep pattern of ordered relationships. Where does this order come from? Let us see if we can find a clue to this in the way that organisms are made.

We can start by asking what makes whorls such a stable, robust morphological character of the whole order of giant unicellular green algae, the *Dasycladales*. There is great variety of form in the whorls and laterals of different species, as you can see in Figure 4.14 and the species shown in Figures 4.5 and 4.6. These are all variations on a common theme. In some species the theme is masked in the adult. Look at Figure 4.15. The adult form of this species, *Bornetella sphaerica*, has a spherical shape with a faceted cortex (surface) made up of close-packed hexagonal, pentagonal, and heptagonal units.

Where are the whorls? Look at the young cells of this species (Figure 4.15) and you can see them. The whorled laterals undergo a surprising modification as the alga matures, expanding at their tips to form the beautiful close-packed structure of the adult form. Another variant on this modification, *Cymopolia van bosseae,* is shown in Figure 4.16; the close-packed structure is repeated periodically along the axis. This form is made up of two wavelengths, the shorter one of the repeating whorls and a longer, superimposed wavelength of the close-packed structure giving the repeating beadlike arrangement. Who would have believed that these are all single-celled organisms? And why is the whorled theme such an invariant feature of the order? It is this type of stable, persistent structure in a group of species that makes possible a systematic classification of the group, based on shared characteristics and variants of these. What is the reason for the stability of the basic form?

Figure 4.14   *The array of whorls on the species* Acetabularia caliculus
prior to and after cap formation.

(a)

(b)

(c)

(d)

Figure 4.15   *The adult form of the species* Bornetella sphaerica *shown from the exterior (b and c) and in section (d) gives no indication of whorls of laterals. However, the juvenile form of the species shows these very clearly (a). The adult form arises by a remarkable transformation of this into a sphere.*

109

Figure 4.16   *The species* Cymopolia van bosseae *is a further elaboration of the* Bornetella *structure, the close-packed form repeating along the axis.*

In our computer simulations, we studied the patterns of calcium and mechanical strain that could occur either in a sphere, representing the zygote, or in a hemisphere, representing the first stage of the regenerative process. At first we examined what patterns could arise without growth. As we varied parameter values, every form that an excitable medium can generate was revealed: spatial uniformity, gradients, stationary waves, propagating waves, and chaos. The propagating waves included spirals that rotated around the hemisphere. In all of these, there was no evidence of any preferred pattern. However, when we allowed growth, there was an immediate stabilization of a particular growth mode, prominent among which was the sequence of gradient formation at the tip; growth; annulus formation; and then the whorl pattern. It seemed that the growth process was itself a major stabilizer of this sequence. An excitable cytoplasm bifurcates to an initial gradient that is stabilized by plastic deformations of the tip, whose growth then makes possible the appearance of the next mode

110

of a pattern sequence, an annulus. The shape then changes again with tip flattening, creating the conditions for a further bifurcation, breaking the circular symmetry of the annulus to produce the whorl pre-pattern, which is again stabilized by growth at each of the calcium peaks to produce the laterals—which could only start in our model, growth failing on this scale. Each lateral is basically a repeat of tip morphogenesis, involving growth and then a further bifurcation to even smaller elements as the lateral extends. This produces a kind of self-similarity of structure on successively smaller scales, as in a fractal pattern, which is quite common in the growth of higher plants also. It seems that the cascade of symmetry-breaking bifurcations that forms the basic pattern generator of the *Dasycladales* structure is a result of both excitable dynamics *and* change of shape, the two together resulting in the robust sequence producing the morphological theme that gives this group a taxonomic unity. Therefore, the distinctive quality of morphogenetic dynamics in living organisms appears to be shape generated within and by a moving boundary, the dynamics changing the shape while the changing shape feeds back into the dynamics, stabilizing the modes that generate the form and creating conditions for the next bifurcation. Ordered complexity therefore emerges through a self-stabilizing cascade of symmetry-breaking bifurcations that have an intrinsically hierarchical property, finer spatial detail emerging within already established structure, as whorls arise from tips and fine branchings occur in growing laterals. These hierarchical cascades of bifurcations are a characteristic feature of morphogenesis in all species, as we shall see in the next chapter.

The point I want to emphasize now is the intrinsic stability of such cascades when they are linked to the changing shape of the developing organism. From this perspective, the *Dasycladales* constitute a natural group not because of their history but because of the way their basic structure is generated. The historical sequence in which different species have evolved is of considerable interest, for it may tell us something about their neighborhood relations in the space of parameters (genes) that define the domains leading to different forms. But we can make sense of this history only within the context of a morphogenetic theory describing how the different forms are generated. This is a theory of what Stephen Jay Gould has called morphospace, the space of possible morphologies for species organized according to certain principles. Our theory suggests that whorls of laterals arising from an unbranched stalk are typical of the order, because this is an intrinsically robust, generic form that the morphogenetic field of this type of organism generates.

An image of this is presented in Figure 4.17. Here morphospace is represented as a two-dimensional parameter space, though in reality it has many more dimensions than this. In our simplified description

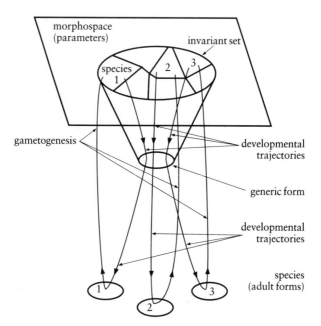

Figure 4.17   *The relationships among morphospace, invariant sets, morphogenetic trajectories, generic forms, and species life cycles. Parameter domains in morphospace described as species 1, 2, and 3 result in morphogenetic trajectories that converge on the generic form and then diverge to produce species morphologies (adult forms). Gametogenesis leads back to morphospace, completing a viable life cycle if the parameters fall within the invariant subset of the species. The invariant set for the whole group, identified and defined by the generic form, is the domain of parameters that includes all the viable species that share the generic form.*

it is the space of genes and environmental influences that specify which morphogenetic trajectory the organism will follow. The large elliptical domain labeled *invariant set* represents a region in which all parameter values produce trajectories that lead to a generic form, described as the narrow neck of the cone through which these trajectories pass. This represents a structure that is typical of a whole group of species, describing a high-level taxonomic characteristic that all the members of this group share. For the *Dasycladales,* this is a cell with whorls of laterals. Although species such as *Bornetella sphaerica* (Figure 4.15) and *Cymopolia van bosseae* (Figure 4.16) look quite different in their adult forms from this basic structure, they pass through it as juveniles and then undergo secondary modification later. This radiation of secondary forms is shown in Figure 4.17 as the diverging arrows leading to the different species 1, 2, and 3. These originate from parameter domains labeled 1, 2, and 3 in morphospace. They all lie within the invariant set leading to the generic form.

During reproduction, each species produces gametes with genes defining parameters that specify what morphogenetic trajectory the zygote will follow. So gametogenesis leads back to morphospace, as shown by the upward trajectories. This process completes the life cycle. Each species has a closed loop, with a range of tolerance for genetic variation that defines the viability limits of the species. The whole group lies within a set of tolerance parameters that are called *invariant* because they result in viable life cycles that return, via gametogenesis, to the same set. Of course, random variations can result in trajectories that fall outside the set, giving either nonviable organisms or transients leading to other invariant sets (other taxonomic groups). The domains of species are invariant subsets, and there are other partitions of these sets that define genera, families, and so on, corresponding to hierarchically related invariant subsets. (I am indebted to Peter Saunders, Department of Mathematics, King's College London, for suggesting this use of the term invariant set.) Here is an ambitious project that is now a possibility: to generate by computer simulation the range of forms represented within the order of unicellular green algae using an extended version of the model I have described, and to explore the sizes of the different basins of attraction of the various genera, and possibly even species, that can arise. Some of these might turn out to be possible forms that are not represented in either fossil or living species.

What this set of simulated algal forms would describe is the morphospace of the *Dasycladales*. Within that space, different morphological forms, different taxa, could be related to one another by a distance measured in parameter space: how much change is required in the parameters to get from one taxon to another. Relationships of similarity measured in these terms would then define a logical or a rational taxonomy of forms based on their generative dynamics. This would give us, within a small corner of taxonomy, a theory of biological forms whose equivalent in physics is the periodic table of the elements, constructed on the basis of a theory that tells us the dynamically stable patterns of electrons, protons, and neutrons. Biology would begin to look a little more like physics in having a theory of *organisms* as dynamically robust entities that are natural kinds, not simply historical accidents that survived for a period of time. This would transform biology from a purely historical science to one with a logical, dynamic foundation. However, there are good reasons to believe that no such scheme of biological forms could ever produce a complete description of all possible organisms, even within a particular order such as the *Dasycladales*. This is because there is a radical unpredictability in the dynamics of these nonlinear systems, which are always open to unexpected novelty. This is the creative foundation of biological process, which expresses itself at the level of structure in the extraordinarily diverse and varied morphologies of species. So a rational taxonomy of biological forms could do no more than reveal the logical order behind the major themes of the evolutionary drama as it has been played out so far. The future will always contain revelations. Nevertheless, the rational and intelligible foundations of the living adventure can be greatly extended beyond their present boundaries. In the next chapter, we explore the idea of evolution as the emergence of generic forms.

The Evolution of Generic Forms

What is it that we see played out in the drama of evolution? What is revealed in the panoply of forms that make such an absorbing spectacle for us to contemplate, from the simplest of cells to the myriad types of marine life, on to giant ferns and dinosaurs, and thence to mammals and humans? Obviously, this tapestry of living process can be interpreted in many different ways. Darwinism stresses the accidents of history; random shuffling of the genetic pack; competitive interactions between individuals for scarce resources; and the power of natural selection to prune out the weak and the unadapted, leaving those that are fit for further reproduction and the perpetuation of their superior genetic legacy. Evolution is seen as struggle, and the focus is on the individual. Even species are seen as individuals in the sense that they are the products of historical accident and the necessities of survival. One of the most eminent contemporary philosophers of science, David Hull, has dedicated his talents to the clarification of this basic Darwinian concept of the species as a historical individual, the result of accident and contingency (see "Historical Entities and Historical Narratives" in *Minds, Machines, and Evolution*). From this perspective, classifications of species, or biological taxonomies, tell a purely historical tale, the accidental adventures of life, which in neo-Darwinism has come to mean essentially the adventures of genes. That is the meaning of evolution.

History, accident, and survival are certainly part of the evolutionary story, as they are of any process, animate or inanimate. However, we saw in the preceding chapter that these do not provide a satisfactory basis for understanding a biological form such as *Acetabularia*, whose whorls do not appear to be there by accident, nor do they in any obvious sense seem necessary for its survival. *Acetabularia* does not seem to be maximizing its fitness, and yet it has been a very successful survivor over millions of years. Why doesn't this species get rid of whorls altogether and put its resources into reproduction—make more caps, for example? The conclusion of the preceding chapter is that morphogenesis is a process that has intrinsic properties of dynamic order so that particular forms are produced when the system is organized in particular ways. This is the lesson about natural order that we learn from physics. Why should biology be any different? *Because it is so complex,* it has been argued. But out of complexity, nonlinear dynamics, robust order emerges. That is the theme I want to develop now, in the context of evolution: there is an inherent rationality to life that makes it intelligible at a much deeper level than functional utility and historical accident.

Patterns in Green

To illustrate this argument I start with another example from the world of plants. There are about 250,000 different species of higher plants—the ones that are familiar to us, with roots and stems, green leaves, and flowers. The detailed structure of the leaves, and the shapes, sizes, and colors of the flowers in these different species have an unimaginable diversity, a stunningly beautiful spectacle of inventive variety in the plant world. However, underlying this diversity is an unexpected and startling degree of order. Despite the profusion of leaf shapes in higher plants, there are basically only three ways in which leaves are arranged on a stem. One way is for them to occur one at a time at fairly regular intervals, with nearest neighbors on opposite sides of the stem. A familiar example is maize (corn), shown in Figure 5.1 (right). This is typical of grass species, technically called *monocotyledons* because the first embryonic leaf (cotyledon) of a seedling, which serves as a storage organ with nutrients for the developing plant, occurs as a single structure. This alternate arrangement of leaves, resulting in a line of leaves up each side of the stem, is given the name *distichous phyllotaxis* (distichous, from Greek, means having two lines or rows; phyllous means leaf; taxis, order). Another pattern is a whorl of two or more leaves at one position (node) of the stem, and at the next node a whorl with the same number of leaves but with the leaf positions rotated so that they are located over the gaps of the previous whorl. An example (Figure 5.1, middle) is *Fuchsia,* with two leaves in a whorl (*decussate phyllotaxis*). The third, and most common, pattern is *spiral*

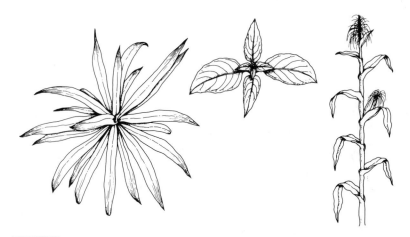

Figure 5.1   *The three types of phyllotaxis: spiral as in yucca (left), decussate as in* Fuchsia *(middle), and distichous as in maize (right).*

phyllotaxis, successive leaves on a stem being located at a fixed angle of rotation relative to the previous one, measured in one direction (clockwise or counterclockwise looking down on the plant). Figure 5.1 shows this pattern in a yucca (left). What is most remarkable about the spiral pattern is that, irrespective of the species, the rotation angle tends to have one of only a few values, the most common one being 137.5°. Figure 5.2 is a schematic diagram that shows the positions of successive leaves relative to one another at the tip of a plant, and the way the angle between them is measured. The growing tip of a plant, or the *meristem,* is a dome that is very similar in size and shape to the tip of a developing *Acetabularia,* but it is made of many cells. A photograph of this structure is shown in Figure 5.3. This is a meristem from a sweet gum tree that grows on the campus of Stanford University near the laboratory of Paul Green, who has some very interesting ideas about the origins of such extraordinary regularity in the patterns of leaf arrangement. What you can see in Figure 5.3 is the sequence of leaves as they are produced on the meristem, the youngest leaf (10) being nothing but a little bump, with older leaves getting larger and larger back to the first or oldest leaf (1) in this picture. Leaf 2 was removed in order to see the others. There are species of plant, such as the family of Bromeliads, whose leaves are arranged in spirals but

divergence angle 137.5°

1. Distichous (corn)   2. Whorled (maple, mint)   3. Spiral (ivy, lupin, potato)

Figure 5.2   *Different phyllotaxis patterns viewed from the top, showing how the divergence angle is measured.*

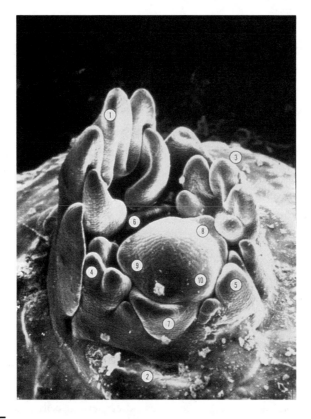

Figure 5.3   *Meristem at the tip of a sweet gum tree, showing the positions
of successive leaves.*

whose flowers have a distichous pattern of petals; and it often happens
that leaf arrangement will start off with a distichous pattern and then
transform to spiral. So it appears that the different patterns are not
fixed characteristics of different species but are a set of alternative
states available to the leaf-generating process in the meristem. As Paul
Green puts it, the problem is not to devise three separate types of
engine for the different patterns, but one engine that can operate in
any of the three modes and switch among them. The question is: What
is the nature of the generator in the meristem that has these properties
and unifies the different patterns as transformations of one another?

It appears that a process is at work that is very similar to that described for *Acetabularia,* whose whorled pattern of laterals is akin to leaf whorls, though there are usually fewer leaves in a whorl of a higher plant than there are laterals in the alga. The research of Paul Green and his colleagues points strongly in the direction of a morphogenetic field in the meristem, defined primarily by the mechanical strains in the surface layer of epidermal cells acting as an elastic shell that resists the pressure exerted by the growing tissue underneath. Epidermal cells respond to this pressure by synthesizing cellulose microfibrils and arranging them in their cell walls in directions that resist the outward-pushing force. This is a perfectly natural process. If you take hold of your sweater with both hands and stretch the wool, the fibers get aligned along the lines of tension or stress, resisting it. In living tissue, new fibers are produced, and they naturally get aligned in parallel with the ones already present, reinforcing the capacity to resist the stress. As leaf primordia grow in a meristem, they create lines of tension in other regions of the tip, depending on the pattern of growing leaves. A single leaf growing on one side of the meristem, as in maize, will create stress lines on epidermal cells located at the surface on the opposite side. Cellulose reinforcement is produced and laid down in cells along these lines that run laterally on the meristem. These cells resist the pressure from growth of the underlying tissue better along the direction of cellulose alignment than at right angles to this, so the tissue will buckle laterally and fold outward under the pressure, producing a leaf primordium on the opposite side to the closest growing leaf. Green has produced a model that describes these forces and has shown how they result in the basic patterns of phyllotaxis. He has also shown experimentally that the cellulose microfibrils behave as expected. What he has not shown is that the only stable patterns of leaf arrangement are the three types shown in Figure 5.2.

Do you notice anything in this figure that you have seen before? Look back at Figures 3.2 and 3.5, showing concentric circles and spirals in the Beloussov-Zhabotinsky reaction and in the cellular slime mold amoebas. It does not take much imagination to recognize that the distichous and the whorled patterns are patterns based on concentric circles, while the other pattern is the familiar spiral form. The difference between examples of patterns in excitable media and patterns in the plant meristem is that the latter are expressed in terms of discrete elements, such as leaves or the parts of the flower. Otherwise, we see the same predominant patterns emerging. So is the meristem also an excitable medium? There are good reasons to believe so.

Fibonacci and the Golden Section

The discreteness of the elements in leaf and flower patterns has some fascinating geometrical consequences that have given rise to a remarkable body of mathematical results connecting leaf patterns with one of the most ancient principles of structural proportions, discovered and used by the Greeks in their architecture. Figure 5.4 shows the arrangement and sizes of successive leaves at the tips of three branches of different size in a single plant, *Araucaria excelsa,* a species of Monkey Puzzle tree. The figure originally comes from a book by A. H. Church, *On the Relation of Phyllotaxis to Mechanical Laws,* published in London in 1904. Successive leaves have been numbered in reverse order to that for the sweet gum tree, starting with the latest leaf (1) in the sequence. The spirals are drawn through leaves that make contact with one another. These are called *parastichies,* and there are always two main spirals of this type, running in opposite directions as shown. Now look at the number series along any such parastichy in Figure 5.4(a). One of these (starting at leaf 1) goes 1, 9, 17, 25, 33, 41, and so on. The difference between each successive number is eight. Now look at the spiral in the opposite direction—1, 14, 27, 40, 53, . . . — the difference is thirteen. This pattern is called (8,13) phyllotaxy. Figure 5.4(b and c) shows the patterns at the tips of smaller branches. The number differences in Figure 5.4(b) for the two spirals are five and eight, so this is (5,8) phyllotaxy; while in Figure 5.4(c) they are three and five, giving (3,5) phyllotaxy. Notice that in Figure 5.4(c) the larger difference (five) is on the counterclockwise spiral (using the

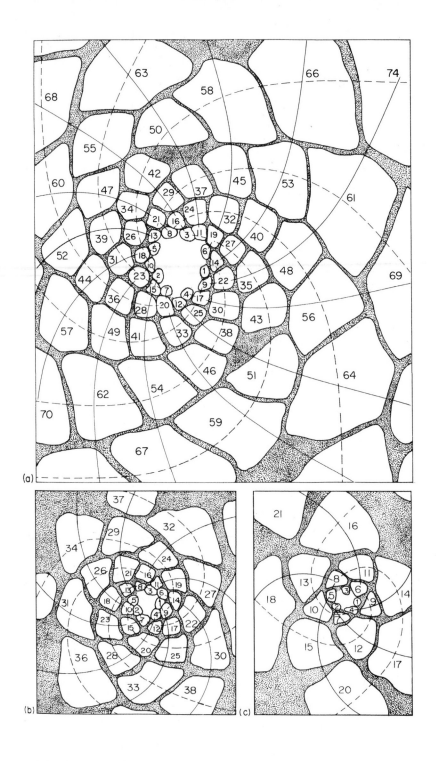

Figure 5.4 *Leaf arrangements on branches of a species of Monkey Puzzle tree,* Araucaria excelsa.

convention that the spiral runs from the center outward). The larger difference (thirteen) in Figure 5.4(a) is also counterclockwise. These define the steeper spirals of the pair. In Figure 5.4(b), on the other hand, the larger difference (eight), defining the steeper spiral, is clockwise. So these directions are not fixed by the genes in this species, since a single plant can have the steeper spiral running in either direction in different branches. Some small initial stimulus initiates the direction of the spiral in any meristem, just as some initial condition in the water of the bathtub breaks the symmetry of the field and starts the flow one way or the other. There are other species in which gene products themselves break the symmetry in one direction, so that all the spirals run the same way in a plant.

Returning to the different number pairs that characterize spirals, comparison of different species of plants reveals a remarkable result: in the great majority of cases, the number pairs belong to a mathematical series that was defined in the thirteenth century by the Florentine mathematician Leonardo Fibonacci. The Fibonacci series takes the form

$$x_{n+1} = x_n + x_{n-1}$$

Any member of the series is the sum of the previous two numbers. If we start with $x_0 = 0$ and $x_1 = 1$, then the series takes the form

$$x_2 = x_1 + x_0 = 1 + 0 = 1$$
$$x_3 = x_2 + x_1 = 1 + 1 = 2$$
$$x_4 = x_3 + x_2 = 2 + 1 = 3$$
$$x_5 = x_4 + x_3 = 3 + 2 = 5$$
$$x_6 = x_5 + x_4 = 5 + 3 = 8$$
$$x_7 = x_6 + x_5 = 8 + 5 = 13$$

and so on. Fibonacci derived this sequence to describe the number of rabbits expected after n generations (starting with two and making certain assumptions about reproductive rate, survival, and death), showing precisely what it means to breed like rabbits. He certainly wasn't thinking about plant leaf patterns. However, by one of those extraordinary coincidences that occur in mathematics, his series generates numbers whose successive pairs define the phyllotaxy patterns of leaves on a plant. Successive pairs of the series describe most of the spirals observed in any species, such as the sunflower (Figure 5.5), the pineapple, or various types of pinecone, as you can confirm for yourself. Now why should this be?

Mathematicians love to play with numbers and discover patterns. Here is one of the properties that was discovered about the Fibonacci series. Take successive members of the series and calculate their ratios:

$$x_2/x_3 = 1/2 = 0.5$$
$$x_3/x_4 = 2/3 = 0.66$$
$$x_4/x_5 = 3/5 = 0.60$$
$$x_5/x_6 = 5/8 = 0.625$$
$$x_6/x_7 = 8/13 = 0.615$$
$$x_7/x_8 = 13/21 = 0.619.$$

Figure 5.5   *Spiral patterns in the seeds of a sunflower.*

124

The series wobbles, or oscillates, up and down, the differences between successive numbers getting smaller and smaller and converging to 0.618 (rounded off). This happens to be a very interesting number that has been known for a long time. Suppose you want to subdivide a rectangle into a square and a smaller rectangle such that the smaller rectangle has the same proportions as the initial one. What should these proportions be? Take the larger side of the big rectangle to be length 1 (for convenience) and the smaller side to be length a, as shown in Figure 5.6. Now a will be the larger side of the small rectangle, and let its smaller side have length b. We want the proportions to be such that

$$1/a = a/b \qquad (5.1)$$

so that the sides of the two rectangles have the same ratio to one another. Now $a + b = 1$, or $b = 1-a$. So equation (5.1) can be written as

$$1/a = a/(1-a)$$

which can be rearranged to give the quadratic equation

$$a^2 + a - 1 = 0.$$

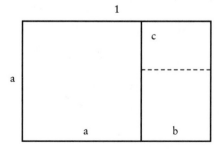

Figure 5.6   *A larger rectangle divided into a square and a smaller rect-angle that has the same ratio as the larger rectangle.*

What is the positive root of this quadratic? The answer is

$$a = [-1 + (5)^{1/2}]/2 = 0.618!$$

The Greeks called this the Golden Section or the Golden Ratio. They based their temples and other buildings on rectangles with this ratio because it allowed for successive subdivisions of the building into squares and rectangles that were similar to the original one. The smaller rectangle in Figure 5.6 can be subdivided further into a square b and a rectangle of sides b and c having the same ratio of 0.618, and so on. This defines a self-similar sequence of forms of progressively smaller size.

Now let's apply the same reasoning to another geometrical figure: a circle. We want to divide the perimeter of the circle shown in Figure 5.7 into two parts, a and b, such that

$$1/a = a/b$$

where the perimeter of the circle is taken as $1 = a + b$. When this is solved for a it is clear that it will give the same value as before,

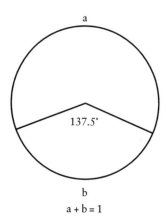

Figure 5.7   *The perimeter of a circle divided into two parts, a and b, whose ratio is the same as that of the whole circle to a.*

since it is defined by the same equation. This then allows us to determine the size of the angle at the center of the figure. We can get this from $b = 1-a = 1 - 0.6180339 = 0.3819661$ (to be exact). The angle is $0.3819661 \times 360° = 137.5°$! So plants with spiral phyllotaxis tend to locate successive leaves at an angle that divides the circle of the meristem in the proportions of the Golden Section. Plants seem to know a lot about harmonious properties and architectural principles. How are we to explain this?

A recent model of phyllotaxis by French scientists Stéphane Douady and Yves Couder gives a significant insight into this question. They used a simple physical model to generate patterns by letting drops of a ferrofluid (a fluid with magnetic properties) fall at the center of a disk covered with a film of oil, in which the drops floated. A magnetic field polarized the drops so that they became little magnetic dipoles that repelled each other. The morphogenetic field of the meristem was thus represented by a magnetic field (see Figure 5.8). As the drops fell at the center of the disk, they experienced a repulsion from polarized drops already present, and they also were exposed to a steady magnetic field that pushed them out from the center toward the edge of the disk. The result is that different patterns arise depending on the conditions of the experiment, but they all correspond to observed phyllotaxis patterns.

If the drops are added slowly, then by the time the next drop is added, the only drop that has any influence on it is the immediately preceding one; the others are too far away to have any effect. As a result, a new drop is repelled to a position 180° away from the previous one, so that the pattern produced is like alternate or distichous phyllotaxis, as in maize (Figure 5.1). As the rate of adding drops (equivalent to the rate of initiation of leaves in a meristem) is increased, a new drop experiences repulsive forces from more than one previous drop, and the pattern changes: the initial simple symmetry of the alternate mode gets broken, and a spiral pattern begins to appear. It takes a while for the system to settle on a steady pattern, the duration of this transient depending on the rate of adding drops. If this is rapid, so

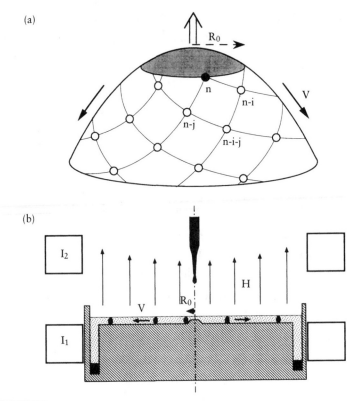

Figure 5.8 (a) Meristem of a plant, showing the positions of leaves and their relative numbers in spiral phyllotaxis, where i and j are successive pairs of Fibonacci numbers; (b) Experimental apparatus to simulate leaf patterns.

that there is strong interaction between drops, then a stable pattern emerges rapidly and successive drops quickly settle into a divergence angle of 137.5°, the spirals obeying the normal Fibonacci series. An example of this is shown in Figure 5.9, which also shows the rapid convergence of the divergence angle on the value for the Golden Section, starting at 180°, which is the angle that will always form between the first two drops. The actual Fibonacci number pair that describes the spirals depends on the rate of addition of drops, the one shown having the values (13,21), thirteen spiral arms going one way, twenty-one the other. Douady and Couder found other divergence angles for

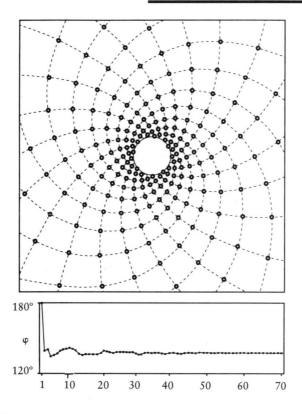

Figure 5.9   *The spiral pattern obtained in which* i = 13 *and* j = 21, *after a transient in which the divergence angle changed as shown in the bottom graph.*

different rates of addition of drops, such as 99.502°, 77.955°, and 151.135°, which are among the minority classes that are occasionally observed in plants. However, they represent much less stable spirals than that generated by 137.5°, which is the only direct symmetry-breaking bifurcation from an initial alternate pattern to a spiral. That is to say, any system that starts with an initial pattern of alternate-leaf primordia (which is the arrangement that the two initial leaves on a meristem tend to start with) will naturally follow a bifurcation to the dominant spiral with a divergence angle of 137.5° if the rate of leaf formation is above a critical value. All other spirals are minority classes that are less robust than the dominant one.

This important result was established by Douady and Couder in the following way. The quantity they used as the parameter controlling the transition from alternate to spiral phyllotaxis is a dimensionless number that they define as $G = V T/R_o$, where V is the rate at which drops are moving away from the center of the disk under the action of a steady magnetic field, T is the period between addition of drops, and R_o is the radius of the region that corresponds to the center of the meristem around which the leaf primordia are generated (see Figure 5.3). As the rate of addition of drops (rate of leaf formation) increases, T decreases, and so the transition from alternate to spiral phyllotaxis occurs when G is *less* than a critical value, which they designate as $G_{1,1}$. The (1,1) here is the first pair of numbers in the Fibonacci series (1,1,2,3,5,8, ...) and corresponds to alternate phyllotaxis (one spiral only joins successive leaves or drops, and this spiral can be drawn in either direction, clockwise or counterclockwise). For values of G less than $G_{1,1}$, spiral phyllotaxis begins, and as G is decreased continuously, the normal sequence of Fibonacci spirals is generated, corresponding to successive pairs of numbers, with fairly abrupt transitions between them. The divergence angles corresponding to successive values of G are shown in Figure 5.10. The main curve starting at $\phi = 180°$ (triangles) converges toward $\phi = 137.5°$, with oscillations about this value as G decreases and the phyllotaxy number changes systematically. Two series are shown, corresponding to different energy functions describing the strength of the inhibition between the drops ("leaves"). The exact values of $G_{1,1}$ at the transition from alternate to spiral phyllotaxis differ, and so do the details of the curves, but their basic properties remain unchanged; that is, the model is robust to differences in the details of the morphogenetic field. However, there is a quantitative relation that does not change, and it is this that establishes the transition as a bifurcation. Douady and Couder show that the divergence angle varies as

$$180 - \phi = (G_{1,1} - G)^{1/2}$$

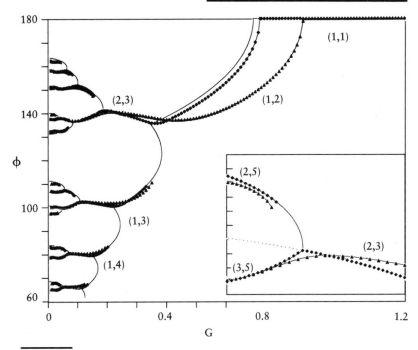

Figure 5.10
How the divergence angle changes as the parameter G varies from 1.2 to a small value. The main bifurcation sequence converges to the divergence angle of 137.5°, while other divergence angles are obtained as minority classes. Two different energy functions were used to show that the convergence to 137.5° is independent of interaction energy. The insert shows a transition to a minor branch of the bifurcation in detail.

a parabolic relationship between φ and G in the neighborhood of the transition. This identifies it as a symmetry-breaking bifurcation. The implication is that the robust sequence for plant meristems is precisely the major Fibonacci series that is observed in higher plants: the arrangements of leaves observed in nature are the generic forms that result from a self-organizing, robust morphogenetic process. The angles other than φ = 137.5° can occur under particular circumstances, and are occasionally observed in nature, but they are minority classes that are not reached by the main sequence. So plants generate this aspect of their form simply by doing what comes naturally—following robust morphogenetic pathways to generic forms.

There is one pattern missing from this description: whorled phyllotaxis. This would be obtained if more than one drop were added at any one time, with $G > G_{1,1}$ so that only the influence of the previous cluster of drops were experienced. Then each cluster of drops would take up positions at maximal distance from each other, and successive clusters would be arranged in the spaces of the previous cluster, resulting in a whorled pattern. So all the patterns can be generated simply by changing growth rates and numbers of leaves generated at any time; these are presumably the main parameters that differ among plant species. The neighborhood relations of the different members of the sequence are clearly defined by their proximities in the symmetry-breaking process described in Figure 5.10 and the transitions to the possible minority Fibonacci number pairs. These are all transformations of one another under changes of parameters and initial conditions.

Over 80 percent of the 250,000 or so species of higher plants have spiral phyllotaxis. This is also the dominant form generated in the model, which identifies it as the most probable form in the generative space of possible phyllotactic patterns. So we get an interesting conjecture: the frequency of the different phyllotactic patterns in nature may simply reflect the relative probabilities of the morphogenetic trajectories of the various forms and have little to do with natural selection. That is to say, all the phyllotactic patterns may serve well enough for light-gathering by leaves and so are selectively neutral. Then it is the size of the domains in the generative space of these generic forms that determines their differential abundance. This is not to deny that the forms taken by organisms and their parts contribute to the stability of their life cycles in particular habitats, which is what is addressed by natural selection. It is simply to note that an analysis of this dynamic stability of life cycles can never be complete without an understanding of the generative dynamics that *produces* organisms of particular forms, because their intrinsic stability properties may play a dominant role in determining their abundance and their persistence. The objective is not to separate these different aspects of life cycles, but to unify them in a dynamic analysis that puts natural selection into its proper context: it is in no sense a generator of biological form, but it may be involved in testing the stability of the form.

Genes and Genericity

The patterns of phyllotaxy in higher plants are clearly candidates for generic biological forms—naturally stable states of a generative process in the developing organism, in this case in the meristem. What role do genes play in producing these? As suggested in the preceding chapter, genes define the region of parameter space where a particular species starts its development. This is determined by such quantities as the turgor pressure within the cells of the meristem, the mechanical properties of the cellulose microfibrils, the composition of the cell walls, the activities of pumps and channels that regulate concentrations of ions such as calcium, and a host of other properties that have been extensively studied by plant physiologists. As development proceeds, these and other quantities affecting the morphogenetic field in the meristem, including plant hormones such as auxins and kinetins that influence rates of growth, will influence the trajectory that the developing system follows. According to the model of Douady and Couder, by simply influencing rate of leaf primordium development the genes can shift the meristem from a distichous to a spiral pattern, or the other way around, as occurs in the transition from spiral leaf phyllotaxis to distichous flower arrangement in Bromeliads. But the genes appear to be working with generic forms, transforming one into another, which is to be expected since the stable, robust modes of the system are the ones that are going to predominate. Work in recent years on the way genes act to influence the patterns produced by the meristem, and the manner in which they work in combination to produce transformations of structural elements are consistent with this view. This work has concentrated largely on the study of how genes influence flower formation.

Geneticists select species according to their convenience for laboratory work—for ease of maintaining stocks, for short generation times, and for a variety of characteristics affected by genetic mutations.

It would not make a lot of sense to study the genetics of the sweet gum or the Monkey Puzzle tree, as an establishment the size of Kew Gardens would be needed to keep large enough populations, and, with generations on the order of twenty to thirty years for such trees, there could be only a few experiments per generation of scientist. So more convenient species are used, and one that has become a favorite among plant geneticists is a little, inconspicuous weed with a tiny flower called *Arabidopsis thaliana*. Despite its size, this plant conforms to the same basic leaf and flower patterns as other species that grow to thousands of times its size. The way genes influence these patterns is beginning to become clearer.

Arabidopsis produces flowers in patterns that actually combine in one structure both spiral and whorled forms. Flower primordia, the beginnings of flowers, arise in a typical phyllotactic spiral looking initially just like the arrangement of leaves, which is also spiral in this species. However, as these flower primordia develop, they each produce a series of structures in whorled array: first a whorl of four sepals (modified leaves), then a whorl of four petals whose positions are located in the gaps of the sepals. The third whorl consists of six stamens, the organs that produce pollen, while the fourth one is two carpels, which fuse to form the gynoecium with a two-chambered ovary, the female part of the flower with the eggs (see Figure 5.11). These individual flowers develop in sequence so that flowers at all stages of development can be seen arranged along the spiral pattern, the oldest flowers being farthest from, and the youngest being closest to, the growing tip. The whole pattern of many individual flowers is

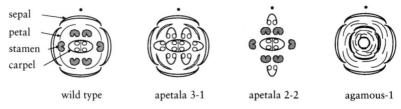

sepal
petal
stamen
carpel

wild type   apetala 3-1   apetala 2-2   agamous-1

Figure 5.11   *Normal and mutant patterns of organs in flowers of* Arabidopsis.

134

called a *raceme,* a structure found in many species and familiar in hollyhock, Canterbury bell, and delphinium.

Mutants of *Arabidopsis* have been discovered that reveal intriguing patterns of variation on this basic theme. In one of these the first whorl of organs in the flower is normal (sepals), but the petals are transformed into sepals as well. Then there is a whorl of six carpels in place of the usual stamens, with normal carpels in the fourth whorl (see *apetala 3-1,* Figure 5.11; *apetala* means "no petals"). There is another mutant phenotype that also has no petals but the transformations now occur in whorls 1 and 2 instead of 2 and 3. The first whorl of sepals is replaced by two carpels, while the petals of the second whorl become stamens, which combine with the normal third whorl, with normal carpels in the fourth. This is called *apetala 2-2.* The third main category of flower mutation affects whorls 3 and 4, transforming stamens into petals and carpels into sepals. So this flower has no stamens or ovary and hence is sterile. The mutant is called *agamous-1* (no gametes). A very intriguing feature of this mutant is that the sequence of whorls of sepals and petals often repeats itself so that the flower consists of sepals-petals-petals-sepals-petals-petals-sepals. It does not seem to know when to stop, rather like a leaf meristem whose normal state is to go on producing leaves indefinitely—hence the capacity of trees to grow to enormous size. The result is a superflower, very decorative but not much good for reproduction. When these three mutant genes are combined in pairs in a single plant, the result is that the four whorls all develop into one type of organ—carpels, or sepals, or organs intermediate between petals and stamens, depending on the combination. And in a triple mutant carrying the three mutations, all the floral organs develop as whorls of leaves.

It has been known for a long time that the different organs of a flower are transformations of one another, and that all are transformations of leaves. This conclusion was based on the observation of intermediate states between organs that occur spontaneously in plants. Just over two hundred years ago, in 1790, Johann Wolfgang von Goethe proposed that all floral organs are derived from the basic leaf state by what he described as different qualities of sap. That was a stunningly accurate piece of deduction especially, one might be tempted to say, for a poet. However, Goethe himself ranked his scientific work significantly above his literary achievements, which themselves have given him the status of a creative genius of the first rank. So what are we to make of his science, which currently tends to occupy the fringes of conventional research, its originality keeping it marginalized? Goethe believed in a science of wholes—the whole plant, the whole organism, or the whole circle of colors in his theory of color experience. But he also believed that these wholes are intrinsically dynamic, undergoing transformation—in accordance with laws or principles, not arbitrarily. So he was an organocentric biologist, and a dynamic one to boot! It is only now that we can begin again to recognize his insights, which involve an aesthetic appreciation of form and quality as much as of dynamic regularity. The ideas I am developing in this book are very much in the Goethean spirit, and I shall say more about this in the last chapter, "A Science of Qualities." But for now we return to the precision and beauty of genetic and molecular studies connected with the transformations that Goethe discussed.

Homeotic Transformations

The changes of shape described in the organs of a flower are examples of homeotic transformations (*homeo* means "similar") because one organ is replaced by another structure that belongs to the same natural set of forms. An elegant model that explains many of the genetic results just described has been proposed. One's first instinct might be to look for a way of arranging gene action spatially in the meristem so that each of the four types of organ is under the influence of a separate gene, for example, concentric circles of gene expression that correspond to each of the whorls of organs in a flower. However, we have to get four distinct whorls from three types of genes, so the genes must be acting together in some way to produce distinct combinations. The mutants actually tell us what these combinations are, since in each type of mutant two adjacent whorls of organs are affected. So we can propose that there are three concentric circles of gene action, each of which covers two whorls. Calling the genes simply A, B, and C, a possible arrangement of their domains of influence in the floral meristem is shown in Figure 5.12. This shows the tissue of the meristem that will produce the different organs as concentric circles, and the pattern of gene expression within these. The combinatorial code for gene influence is then A = sepals, AB = petals, BC = stamens, C = carpels. We can also assume that if none of these genes is active, then leaves are produced; that is, leaves are the "ground state," as Goethe suggested. What happens when one of these genes is absent (mutates)?

If A disappears, then it might be expected that the sequence would become: nothing (leaves), B (a mixed organ, not normally found), BC (stamens), C (carpels). What is observed is not this but the following: carpels, stamens, stamens, carpels, corresponding to the gene sequence C, BC, BC, C. It appears that when A disappears, C spreads out over the region where A would normally be. This type of behavior is often found in gene expression: genes interact with one another, limiting their domains of activity. That is, genes can control one another's activity by inhibitory interactions. We would then expect that if C is

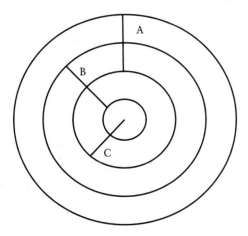

Figure 5.12   *The spatial patterns of the three main categories of homeotic gene involved in determining the structure of the flower organs in* Arabidopsis, *deduced from the mutants.*

absent, A would extend its region of influence across the whole meristem. And this is how the *agamous* mutant seems to work. In the absence of C, the sequence is

A	AB	AB	A
sepals	petals	petals	sepals

So far so good. Now for B. When this mutates to an inactive form, the first expectation is the sequence

A	A	C	C
sepals	sepals	carpels	carpels

And this is what is observed. So there are no complications about spatial patterns changing when B mutates.

As far as it goes, this model, arising from the work done by groups in Pasadena, California, the John Innes Institute in the United Kingdom, and at the Max-Planck Institute in Cologne, Germany, gives a very satisfactory account of these classes of gene activity. These homeotic mutants are a very interesting class of single-gene mutation, and they occur in animals as well as in plants, often with dramatic consequences. For instance, in the fruit fly *Drosophila,* there are homeotic mutants in which parts of the eyes are transformed into legs or wings! And in humans, fingers can be transformed into thumbs and vice versa, a subject that I consider in a later section of this chapter. Homeotic mutants tell us that structures that look quite different—such as legs, wings, and eyes, or petals, stamens, and carpels—are in fact easily transformed, one into another, by the effects of single genes. This gives us very important information about two aspects of morphology: first, that in the space of possible biological forms certain structures are close neighbors from a generative or morphogenetic perspective, even if they have quite different shapes; and second, the clustering of certain forms in regions of shape space suggests why they occur over and over again in different species—such as the four basic whorls of organs in flowers. This is further evidence that there are generic forms in the biological realm, structures that are intrinsically robust. Now we see how such structures can be clustered as neighbors in generative space, readily accessible as basic forms. Homeotic gene activities, which are spatially organized in patterns just like other components of excitable media (as in the concentric circles of gene expression in the floral meristem), stabilize or select particular patterns from the set of generic forms such as the concentric whorls of flower organs. Other genes add distinctive properties such as details of shape, size, and color to particular organs. It is these additional features that result in the prolific variations on basic generic themes, making the realm of plants such a stunning spectacle of creative variety. Genes cooperate with and add variety to generic themes.

Gene Activity Patterns and Morphogenesis

The techniques of molecular analysis are now such that the products of particular genes can be precisely located in the cells of developing tissues. All the evidence currently available points to the correctness of the model proposed for the three groups of homeotic genes that were labeled A, B, and C in the last section (Figure 5.12). Such spatial patterns of gene activity that correlate with morphological changes in mutants have been identified in many developing organisms, the fruit fly *Drosophila* leading the field in providing detailed information about the way gene products are spatially organized in the developing embryo. These gene patterns themselves develop gradually and systematically, having a dynamic of their own. Part of this dynamic arises from interactions between the genes themselves, via their products, such as the inhibition of the activity of gene C by A so that there is a sharp boundary between their domains of action. There are also positive interactions between genes, one gene product activating another so that they have overlapping domains of expression. What has been shown for many of the genes that produce spatial patterns in *Drosophila,* and those that underlie the pattern of floral organs in plants, is that the proteins produced by these genes have a regulatory function due to their capacity to bind to regions of DNA, or to modulate the activity of other proteins with this capacity, so that they either activate or inhibit gene activity. In this respect gene products are just like chemicals that interact and diffuse, producing spatial patterns of gene activity so that we can regard the network of regulatory genes and the distribution of their products as components of the excitable medium of the developing organism. This network consists of many genes in addition to the homeotics, including those that switch them on and initiate the process of flowering in a growing plant. One of these, called *floricuala,* is somehow involved in the plant's response to environmental factors that contribute to the decision to produce flowers instead of leaves. These factors include day length and temperature, added to which there are internal factors such as age and size that influence the timing of flowering.

There is an interesting discrepancy between the spatial patterns of genes shown in Figure 5.12 and the pattern of organs in a flower (Figure 5.11). Whereas the gene products are distributed in continuous rings or circles, the organs are discrete entities, either separated from one another or fused together by a secondary process, as in the carpels. This discreteness of elements evidently comes from the dynamics of the morphogenetic field, not from the genes. We saw in the case of whorl formation in *Acetabularia* how this might come about. The annular ring of calcium that forms and accounts for tip flattening transforms naturally into a series of peaks, the pattern from which a whorl arises. This process is a result of the interaction between the dynamics of the calcium–cytoskeleton field and change of shape in the developing organism. Similar wave patterns around the circular dimension of the floral meristem are likely to arise in the same way by symmetry breaking, the field falling into its stable attractor. Genes naturally do the sensible thing: they cooperate with the generic forms of the field to give robust morphologies to organisms. Genes can influence a lot of secondary properties of these forms, such as the number of elements in a whorl (that is, the wavelength of the pattern); the detailed shapes of the structures produced (petals, stamens, carpels);

their color and perfume; and so on. But generic properties cannot be denied except at a considerable price.

To illustrate this point, consider flowers that appear to violate the principle of discrete elements, such as a snapdragon, or an orchid in which the petal is a closed tube. It turns out that this is a result of secondary fusion of an initial whorl of petals that grow together to form a continuous structure. Evidently, it is easier to modify a whorl than to produce a tubular structure to begin with, even though the pattern of gene expression itself would tend to produce a structure with a circular cross section. For similar reasons, *Acetabularia* finds it easier to produce whorls and discard them rather than to suppress the production of these generic forms. The combination of generic forms and genetic variations on these themes results in both the diversity of form in the biological realm and the intrinsic order of these forms that allows them to be classified. A logical unity is thus revealed that goes beyond the historical unity of Darwin's view of life as a tree of related forms descended from common ancestors, for these ancestral types, unexplained in Darwinism, can now be understood as the generic or typical forms produced by the generative process, the dynamics of morphogenetic fields, that underlies the origins of species. This is not a new idea. It is where modern biology started, with the great eighteenth-century taxonomist Linnaeus, and it is the vision that Goethe had of a dynamic, transformational unity underlying the spectacular diversity of the forms of life. It is the original tradition of biology, conceived as having close affinities with the principles of explanation used in physics and mathematics. Darwin shifted the focus of biology significantly by describing it as a historical science, and twentieth-century biology has added a dimension of molecular and genetic reductionism that has all but wiped out organisms as dynamic forms in transformation, the fundamental units of the evolutionary adventure. But now the new developments in mathematics and the sciences of complexity have revealed ways of reconciling these traditions and linking biology again to the exact sciences. Genes do not control; they cooperate in producing variations on generic themes.

141

Going Out on a Limb

Until now, the discussion about what is revealed in the spectacle of evolving biological forms has been restricted to plants. Now it's time to consider an example that illustrates the same argument in animals. A longtime favorite for examining evolving animal form is the limbs of four-legged vertebrates (tetrapods). Limb bones are very durable and leave clear impressions as fossilized remains in sedimentary rocks, so there is a rich collection of these structures to build a bridge from contemporary species back to the earliest known ancestors that existed some 400 million years ago. Limbs evolved from fish fins, which somehow underwent transformation to the more complex structures that eventually turned out to be so useful for making clay pots, weaving tapestries, and playing the piano. There is a wealth of material here that has been extensively examined and interpreted in different ways according to different views of what this record is telling us.

An extraordinary thing about our limbs is that they are essentially the same as those of all other tetrapods—horses, bats, birds, crocodiles, and those ancient progenitors that lived in the Devonian seas. Just look at the sample of limbs in Figure 5.13 and contemplate their similarities and differences. The first is a hind limb that belonged to a fossil fish, *Ichthyostega*, from the Devonian period. The second is the hind limb of a salamander. Then there are the wings of a bird (chicken) and a bat, the front leg of a horse, and a human arm. All these species have (or had) quite different uses for their limbs: *Ichthyostega* for swimming; the salamander for swimming and walking; the chicken for flapping about; the bat for flying; the horse for a range of gaits between walking and galloping as well as scratching and kicking; and humans for an endless variety of activities. Given this diversity of uses, one might expect that natural selection would have designed each limb to optimally serve its functions. Why doesn't the bat's wing start with two bones to anchor it firmly to the shoulder? Why does the horse have that tiny extra bone running like a splint down the side of its main "toe," with another similar one on the other

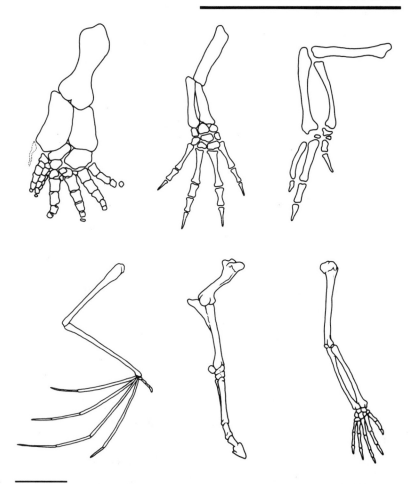

Figure 5.13   *Bone patterns of limbs of various tetrapods:* Ichthyostega, *salamander, chicken, bat, horse, and human.*

side of the toe? What possible function can they serve? Why not get rid of them altogether? Given their extraordinary utility and the fact that *Ichthyostega* had seven, why don't we have six digits on each hand and banish that rather useless little toe that is so prone to getting stubbed? The answers to these questions usually take the following form: Natural selection has to make do with what is given by the ancestral form, molding it as best it can to a variety of purposes. But then we are left with the problem: Where does this ancestral form come from, and why is it as it is? Is it just a historical accident, or is there a deeper reason for the basic pattern of tetrapod limbs that provides a rational unity of structure underneath the diversity of functional expression?

Let's start this inquiry by identifying the basic pattern that is common to the sample of limbs shown in Figure 5.13. They all begin with a single large bone, then two more major bones followed by a cluster of smaller ones that end with the terminal digits, which can vary in number between one (horse) and seven (*Ichthyostega*). Another fossil tetrapod, *Acanthostega,* had eight digits. The number five used to be regarded as the ancestral character so that tetrapod limbs were known as *pentadactyl* (five-digit) limbs. This has now been abandoned, although five has recently returned as a special number in genetic guise, as we shall see.

The basic tetrapod limb pattern, recognized by comparative morphologists since the eighteenth century, was one of the great insights of the pre-Darwinian tradition known as *rational morphology.* One of the most eminent members of this tradition, the French zoologist Étienne Geoffroy Saint-Hilaire, used this structure to exemplify his Principle of Connections, which stated that certain patterns of relationship between structural elements in organisms remain unchanged even if the elements themselves undergo alteration. Thus, the two little bones and the main toe in the horse's limb, which are called metacarpals and are labeled II, III, and IV, have the same relationship to one another and to the cluster of bones above them (the carpals) as counterparts in the other animals' limbs, even though II and IV have become very much reduced in size relative to the highly developed metacarpal III, which articulates with the sequence of three single bones (phalanges) of the one toe that ends in a prominent "toenail," the hoof. Geoffroy regarded all tetrapod limbs as transformations of a single basic ground plan of structural elements so that the diversity of forms is logically grounded in a unity of type. I think you will now recognize the familiarity of this phrase and realize that the position I am arguing for in this book has an old tradition. The difficulty that Geoffroy's concept ran into, however, was that his morphological Principle of Connections was interpreted as a description of a static set of relationships that defined an ideal or archetypal limb from which all others are derived by transformation. One of his most influential disciples, the eminent morphologist Richard Owen, who founded the Natural History Museum in London, was very explicit about these ideal forms, which he regarded as the work of a transcendental creator. Owen believed that the archetype, a kind of Platonic form, takes different shapes in different species according to its use. This was the tradition of morphological explanation that Darwin himself inherited, growing up in Victorian Britain in the early years of the nineteenth century. He knew about Geoffroy's ideas and those of his illustrious compatriot, the Baron Georges Cuvier, who shared with Geoffroy a passion for understanding the biological order revealed by comparative morphology but disagreed with him violently over the extent to which all animal species can be understood as transformations of one another. Instead of Geoffroy's one, Cuvier proposed four basic types of animal: vertebrates, mollusks, arthropods (insects), and radiolaria (for example, sea urchins and hydra). This is not a big difference numerically, but conceptually it is very substantial, and these two Frenchmen battled it out in public debate and private feud. Their intense intellectual conflicts animated the field of comparative morphology, which Goethe followed with the deepest of interest, siding with Geoffroy in seeing all the species of the animal kingdom united with one another under transformation.

Darwin was well aware of these disputes, but he chose another route to unity. His deep interest in historical explanations and his belief in gradual rather than sudden change (species arising by the accumulation of small hereditary differences rather than by discrete jumps) caused him to abandon the search for a logical unity of biological forms and to seek a scheme of genealogical relationships in a tree of life. In this scheme, unity comes from historical continuity, and diversity comes from functional necessity (survival in different habitats). He took Richard Owen's archetypal forms and turned them into common ancestors, the branch points on the tree of life where an accumulation of small variations adds up to a significant difference.

At one stroke he transformed biology from a rational science that sought intrinsic principles of biological order, such as Geoffroy's Principle of Connections in morphology, into a historical science in which virtually any form is possible and the only principle is survival through adaptive modification. This is an extreme position to adopt in science, and it created a sharp separation between the explanatory principles used in biology and those used in other sciences, even though Darwin modeled his evolutionary concepts on the science of geology. This was inspired by the work and writings of the great nineteenth-century geologist Sir Charles Lyell, who maintained that the physical and chemical processes that shaped and transformed the earth in the past are the same as those that operate today. Therefore, we can understand remote events in terms of the principles of physics and chemistry with which we are familiar now—they provide a universal, rational framework for reconstructing the past. Darwin observed the mechanics of change in populations of organisms—domestic animals undergoing selective breeding, for example—and deduced that similar processes shaped species in the past, but that the selection was natural. This is a perfectly rational deduction. But selection has no intrinsic principles, whether artificial or natural; the breeder can select for any trait he or she fancies, and so can nature, "fancies" in this context meaning that which works. So Darwinian biology has no principles that can explain why a structure such as the tetrapod limb arises and is so robust in its basic form: it just appeared in a common ancestor. This leaves a very large hole in biology as an explanatory science and, in many respects, represents a retrograde step from the position of the rational morphologists, who were seeking such principles.

Geoffroy was absolutely right to focus on the relational order of the elements of tetrapod limbs as the key to what is unchanging (invariant) in this pattern. But this order has to be put into a dynamic context that was only beginning to become a systematic and important part of biology in the early years of the nineteenth century. The dynamic context for the understanding of biological form is developmental biology. Goethe recognized its significance and described organisms as dynamic forms in transformation. Darwin was well aware of its importance but saw it as contributing to historical description and the tracing of relationships in his tree of life rather than as a field that could explain his common ancestors as generic forms of the developmental process. The latter is actually closer to Lyell's ideas about geology that inspired Darwin, since it grants that the physics and chemistry of morphogenesis provide the principles of biomorphological explanation, like Lyell's use of physical and chemical principles to explain the changing morphology of the earth. But only in this century have the mathematical tools for this kind of analysis been developed, allowing us to address the issues of invariance, symmetries, and symmetry breaking in complex nonlinear dynamic processes, and giving us insight into the origins of the structural constraints that can explain distinctive features of biological form such as tetrapod limbs. No blame to Darwin for shifting biology onto a different track and sacrificing rational unity for historical unification. There is no reason we cannot have both. Certainly the time is overdue to correct the deficiency and to explore its consequences.

Limb Morphogenesis

A limb of an organism such as a salamander starts off as a little outgrowth from the flank of the developing embryo (Figure 5.14). This is called a limb bud. If this is examined in detail nothing is visible but a lot of cells surrounded by an extracellular matrix filled with a variety of proteins, some fibrous and some globular, and other materials. The cells of the limb bud, called *mesenchyme cells,* have an elaborate cytoskeleton and molecular mechanisms for regulating calcium similar to those described for *Acetabularia* in the preceding chapter. In addition, they have a collection of genes on their chromosomes capable of making a great diversity of proteins, a number of which are involved in limb formation. The extracellular matrix, which the cells produce by secreting proteins of particular kinds and within which they move and interact, is as important in animal morphogenesis as the intra-

Figure 5.14   *The limb bud of a salamander.*

cellular cytoskeleton. In the multicellular plant meristem, cell walls are in intimate contact with one another and constitute a major component of the morphogenetic field, transmitting the mechanical forces and chemical influences that are involved in shaping the elements of plant form such as leaves and flower organs, as we have seen. In the limb bud the extracellular matrix serves a similar role, relaying mechanical and chemical forces across the morphogenetic field of the bud and contributing to the formation of the elements of structure that become the bones. But whereas in the plant, the leaf and flower organ primordia develop on the surface of the meristem, in the limb bud the elements of structure that develop into the bones arise within the bud. How does this happen?

The description that I now present comes from the work of a great variety of experimentalists who all have made substantial contributions to our understanding of limb morphogenesis. These have been integrated into a general theory of limb formation by Neil Shubin and Per Alberch, based on a model of the morphogenetic field of the limb bud developed by George Oster, Jim Murray, and Philip Maini, which I believe provides major insights into its basic properties.

The limb buds form as a result of regions of actively dividing cells in the flank of the embryo that initiate the outgrowths, which are then enhanced by migration of neighboring flank cells into the buds. The zone of dividing cells at the tip of a bud continues to add cells so that it extends out from the flank, taking the form shown in Figure 5.14. At this stage the cells in the interior of the bud all look the same and are distributed within the extracellular matrix with no sign of any differentiated structure. When the bud reaches a particular size, however, the first evidence of spatial pattern arises: cells begin to condense into a more closely packed structure in the middle of the bud. The evidence suggests that this occurs because cells in this central region of the bud begin to secrete the enzyme hyaluronidase, which degrades a principle component of the extracellular matrix, hyaluronic acid. This acid, which reacts with calcium in the matrix to form a salt, calcium hyaluronate, has a strong affinity for water, forming a gel-like material in the limb bud. When calcium hyaluronate is degraded by hyaluronidase, this gel collapses and the cells within it come closer together. Animal cells are constantly exploring their environments by means of little cytoplasmic feelers—filopodia (filamentous feet)—that extend out from the cell. The pseudopodia of amoeboid cells are just larger versions of filopodia. These cytoplasmic extensions that drive cell movement and exploration are expressions of the dynamic activity of the cytoskeleton with its microfilaments and microtubules that are constantly forming and collapsing (polymerizing and depolymerizing), contracting and expanding under the action of calcium and stresses. Cell surfaces are sticky, so when filopodia of different cells encounter one another, they tend to shake hands (feet) and hold on. Then when the filopodia contract, cells are drawn together and stick even more strongly. So a distinct aggregate of mesenchyme cells forms in the central region of the growing limb bud. These condensed cells then produce materials that turn them into cartilage, and later on this becomes bone.

Why do the cells aggregate in the center of the limb bud? Is there some specific chemical signal that is produced here and that tells the cells to start secreting hyaluronidase? But then what initiates the production of this signal in this region? As you can see, it is very easy to get into an infinite regress when you are looking for the initiator of a process. This is why biologists often invoke genes as the repository of all developmental information, the ultimate source of all the instructions for the development of the embryo. But genes can only respond to their immediate biochemical environment in a cell. They do not know where they are in a tissue unless something tells them. So we come back to the morphogenetic field as the source of spatial information. Lewis Wolpert, who has made important contributions to the study of limb formation, calls this "positional information." How is it set up? Recall what happened in the *Acetabularia* model. The regeneration domain of a cell was described as two thin shells (cytoplasm and wall) with an initially uniform pattern of the field variables, calcium and mechanical strain. Because the cytoplasm is an excitable medium, it can spontaneously develop a nonuniform pattern, and the one that formed had a high concentration of calcium (and strain) toward the tip of the regeneration domain. This initiated tip growth and the process of regeneration then followed as a series of spatial transformations and bifurcations, a robust symmetry-breaking cascade.

In the limb bud the whole system of cells embedded in the extracellular matrix, plus the epithelium (the surface layer of cells), makes up the morphogenetic field. This is also an excitable medium that can spontaneously change its state and generate spatially nonuniform patterns, because each of the cells has an excitable cytoplasm, and they communicate with one another mechanically and chemically through the extracellular matrix. So a pattern is bound to arise. The geometry of the limb bud, which is roughly a cylinder with a conical tip, defines the initial symmetries, and one of these is going to get broken. The most obvious one is along the central axis of the cylinder, so that a difference of state develops from the periphery toward the center.

Exactly what variables are involved in this we don't yet know, but something is bound to change that makes the center of the bud different from the periphery, and this starts the whole process of limb morphogenesis. Notice that the initial geometry of the bud, an outgrowth with a roughly cylindrical cross section, is such that we expect there to be a single element produced toward the center, and this is what occurs in all tetrapod limbs: the first element is a single bone, the humerus or the femur. Genes would have to search long and hard to find a region of parameter space where something other than a single element arose first in such limbs. So they don't bother—they sensibly go with the robust, generic form and modify it secondarily to suit various purposes. But why are there two elements in the next part of the tetrapod limb? This is where a detailed model becomes essential to the analysis.

Oster, Murray, and Maini derived equations that describe the processes of osmotic collapse of the hyaluronate gel of the extracellular matrix, contact and contraction of filopods, cell adhesion, and the recruitment of cells to condensation sites by migration along stress lines in the extracellular matrix so that the initial condensation regions grow. They showed that there are two types of pattern change in the growing cluster of cells: division into two branches, making a Y pattern; or separation into two parts along its length. So there are three types of change or bifurcation that can arise in the limb bud tissue.

1. Focal condensation—aggregation of cells to form a tightly packed mass of cells that can grow by recruitment of more cells.
2. Branching bifurcation of the growing condensate.
3. Segmentation of an aggregate into two parts along its length.

These processes are shown schematically in Figure 5.15. The initial single condensation of cells in the developing limb bud undergoes a

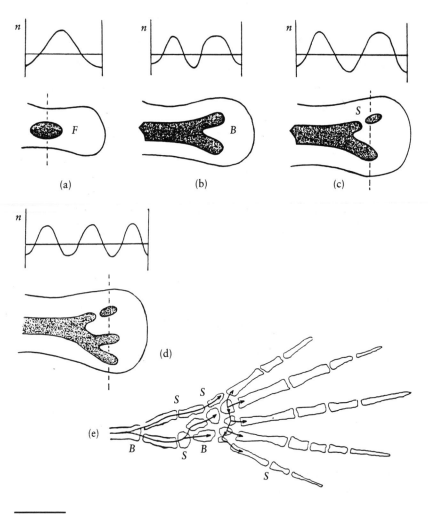

Figure 5.15   *The three ways in which patterns can form in the elements of tetrapod limbs, according to the model of Oster, Murray, and Maini.*

branching bifurcation. Experimental studies show that these branches are initially joined to the first element, as in the Y pattern of Figure 5.15(b), but they subsequently separate. The model shows that it is very difficult to get branching into three elements, so the robust bifurcation is a separation into two, as occurs in all tetrapod limbs, to give the radius and ulna (forelimb) or the tibia and tarsus (hind limb). The profiles of state change across the limbs (dotted lines) are shown above the condensation patterns. After the first three elements have been laid down, there is considerable variation in the pattern among different species. An example of how the model accounts for a particular pattern, that of a salamander, is shown in Figure 5.15(e). The sequence of focal condensations, branching bifurcations, and segmentations is identified by corresponding letters.

Shubin and Alberch have used this model to show how a great range of tetrapod limbs can be produced from the generative rules of the limb field. The problems associated with a historical approach to limb classification can be illustrated by an example from their work. In Darwinism it is assumed that there is historical continuity of limb elements from species to species, with small variations during their evolution, so that the elements can be identified and named as derivatives from a common ancestral limb form. Suppose we apply this to the limbs in Figure 5.16, which belong to two amphibians whose generative sequence has been interpreted as shown. In *Ambystoma*, the fibula (*F*) branches into the intermedium (*i*) and the fibulare (*f*),

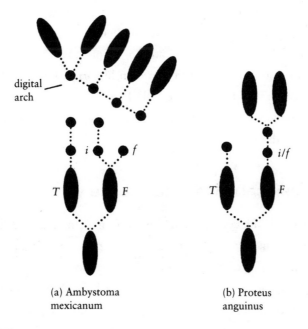

(a) Ambystoma
mexicanum

(b) Proteus
anguinus

Figure 5.16   *Generative sequences of the elements in the limbs of two different amphibians,* Ambystoma *and* Proteus.

and the next elements (the carpals) are initiated by a focal condensation of the elements at the base of the digital arch. In *Proteus,* with only two toes, the fibula does not branch but segments to give a single element. Is this to be named *i* or *f*? Well, it is neither; it's just itself, which can be called *i/f* or anything else. Also, the two digits in *Proteus* arise from a branching of the next element, which segments from *i/f,* whereas in *Ambystoma* the first element of the digital arch arises by a focal condensation. So the digits arise by different processes. One might be tempted to label the two digits in *Proteus* III and IV, and IV and V, but this would be quite arbitrary. Any attempt to identify exact historical lineages for such patterns is bound to run into problems.

Interpretations of historical processes inevitably face difficulties and ambiguities, because we were not there to see what actually happened. On the other hand, we can study what goes on in the embryos of living species. I believe that this provides a much sounder basis for a taxonomy of forms than a genealogically based classification scheme. Of course, the theory that I have been using is going to turn out to be wrong in certain respects, as with all theories, but that doesn't alter the logic of the argument: Generative principles provide a better foundation for understanding structure than historical lineages. Now we need a concept that defines precisely how these structures are to be seen as similar to and different from one another on this basis. We find this in the notion of equivalence.

Classification by Equivalence

Equivalence is the concept that is used in mathematics to define similarity of form. Imagine taking a lump of plastic modeling clay and deforming it according to the following rule: you can do anything to it as long as you don't tear it or make a hole. The set of possible shapes is clearly infinite, but they all preserve a property of the original lump: all parts continue to be connected to one another. As a result, any shape can be transformed into any other shape by simple deformations (no holes or tears). The key concept here is transformation according to some rule, which means that something remains unchanged or invariant despite the change of shape. In the particular case considered, what does not change is the topological property of simple connectedness within the lump of clay that would be violated by a hole or a tear. The possible shapes are all equivalent under transformations that preserve this topological property of connectedness.

There is a hierarchy of geometries defined in this way, forms being equivalent under transformations that keep certain properties of the forms unchanged or invariant. For instance, similarity transformations preserve the angles between the sides of triangles, rectangles, and other figures, but allow their sizes to change. This gives a set of forms that are equivalent under similarity transformations. If distances between points in figures are preserved, then we get metric transformations, which describe the movements of rigid bodies—translations, rotations, inversions. These are among the simplest examples of equivalence of forms under transformations. The concept is capable of great generalization and is crucial in any logical classification scheme. Let's now apply it to tetrapod limbs using the morphogenetic theory described earlier, since what we want is a classification scheme that depends on generative principles.

Tetrapod limbs are defined as the set of possible forms generated by the rules of focal condensation, branching bifurcation, and segmentation in the morphogenetic field of the limb bud. All forms are equivalent under transformations that use only these generative processes. With this we arrive at a logical definition of tetrapod limbs that is independent of history. The idea of a common ancestral form as a special structure occupying a unique branch point on the tree of life ceases to have taxonomic significance. Now tetrapod limbs could have arisen many times independently in different lineages of fish, and they still would be equivalent as long as they were made in the same way, whereas in a Darwinian (historical) taxonomy independent origins mean basic difference. To see how this may have happened, we need to examine how fish fins and tetrapod limbs are related to one another.

Fins are simpler structures than limbs. Their cartilaginous or bony elements are arranged in simple arrays of the type shown in Figure 5.17, with a repetition of the same structure along the length of the fin. These elements are produced by the same type of condensation process as that described for tetrapod limbs, and it is evident that there is a segmentation process that generates the arrays of these elements in the fin. However, there are no branching bifurcations, and there is very little difference of structure from anterior (the head) to posterior (the tail) in the fin, unlike the strong asymmetry that distinguishes the elements across the anterior-to-posterior (A-P) axis of the tetrapod limb and is particularly marked in the human hand with our distinct anterior thumb. So what these structures have in common is that both are produced by focal condensations and segmentations, but tetrapod limbs have in addition branching bifurcations and strong A-P asymmetry. Tetrapod limbs are more complex than fish fins because their production involves these two additional symmetry-breaking processes. The first of these, branching bifurcation, is generic to the production of discrete elements in the mesenchyme of the limb bud, according to the model, but will happen only if the geometry and the dynamics change together in the appropriate way. The limb bud must start off roughly cylindrical in cross section for a single element to start the structure, followed by branching bifurcations that accompany the transition to flattening of the limb in tetrapods. A fin starts as a

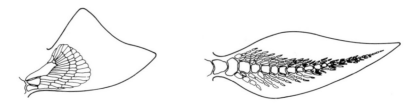

Figure 5.17   *Pattern of the elements in fish fins.*

flat lateral fold of skin so the condensations fit into this geometry. Again we see the intimate interplay between geometry and dynamics, which we encountered in *Acetabularia*, that plays such an important role in stabilizing the cascades of symmetry-breaking bifurcations that result in robust, generic forms.

Fish fins lie within the equivalence class of forms generated by focal condensations and segmentation, which includes the great diversity of possible structures seen in different species. Tetrapod limbs fall within this class as a subset of more complex forms with additional broken symmetries. In this way the relationships of similarity and difference among different classes can be defined, providing a basis for taxonomic (systematic) relationships in terms of generative processes. A rational taxonomy of forms can then define the relationships among the various possibilities that are discretely distributed in morphospace, allowing for the identification of evolutionary trends as changing developmental trajectories of species.

We can identify one of these trends in the evolution of vertebrate appendages. The evidence is that fish fins originated as lateral folds that ran on either side of early vertebrates. These folds then became restricted to two paired appendages, the pectoral and pelvic fins. In the bony fishes, the coelacanths, these became further restricted in lateral extent, resulting in lobed fins that begin the branching bifurcations characteristic of tetrapods, which evolved from the coelacanths. *What we see in this evolutionary trend is a gradual decrease of symmetry and a progressive emergence of increased complexity as more symmetries get broken in the morphogenetic cascade.* This is a natural progression for a dynamical system as its parameters undergo variation by random shuffling of the genetic pack. Any system that starts off simple will tend to get more complex. It has nowhere else to go. Natural selection does not have a lot to do except act as a coarse filter that rejects the utter failures. So we get a description of evolution in terms of dynamics and stability, which always belong together. The question is then: Which are the robust forms that emerge from the evolutionary exploration of the space of possible organisms?

And the answer is: Generic forms arising from the robust symmetry-breaking cascades of morphogenesis that result in species that are stable in some habitat. Fish fins are great for movement in the water. Tetrapod limbs work perfectly well in the water, but they also allow for locomotion on land. So a whole new environment of possibilities is opened up, and more symmetries get broken, resulting in more complex forms, especially in relation to sensory perception and locomotion by tetrapods in terrestrial environments. The path of evolutionary exploration opens up, and more generic forms are discovered by the restless parametric searching of the genes, leading on to us. But we're getting ahead of the story. We need to look at the way genes act in limb morphogenesis, particularly at their role in affecting one of the major broken symmetries: digits. This reveals a surprising and very interesting connection with the way genes act in influencing the sequence of organs in a flower. At the level of morphogenetic mechanisms, animals and plants are remarkably similar.

Homeotic Mutants

Among the set of possible forms of tetrapod limbs are the abnormal structures classified as monsters and mutants. Despite sometimes gross deformation, monsters are nevertheless recognized as belonging to the same set of forms as those classified as normal. William Bateson made very effective use of abnormal forms to make important deductions about the principles of morphogenesis, particularly the significance of symmetries and asymmetries. His volume *Materials for the Study of Variation*, published a hundred years ago, in 1894, is an absolute classic of insight into the deep relations between morphogenesis and evolution of the type described in this book. You can see that there is a well-defined biological tradition of keeping whole organisms and their transformations at the center of the picture, viewing genes as modifiers, not as generators of morphology, and Bateson was one of the most articulate within this tradition. Figure 5.18 is adapted from his book. We have no difficulty in recognizing this as a human limb,

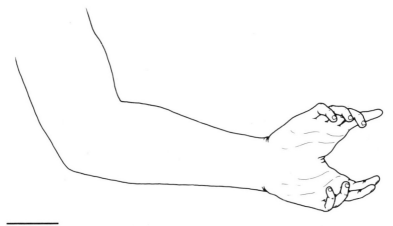

Figure 5.18   *A naturally occurring human hand with a mirror symmetry*
               duplication.

despite the exceptionally large number of digits and the absence of a
thumb. Clearly something has gone wrong with the normal asymmetry
of the hand, and what has replaced it is a structure with mirror symmetry. Such duplications also account for Siamese twins and cases
where there are two or three limbs in place of one. There are always
planes of mirror reflection located between them, just as there is one
between the duplicated hands in Figure 5.18.

In recent years a lot of research on amphibians and chicks has
centered on identifying the molecules involved in these transformations
during limb development, resulting in insights into the way genes
influence these processes. Homeotic mutants have been known for
some time that cause the transformation of one digit into another.

The genes involved in these transformations are clustered together
in groups on chromosomes, and they all share certain properties that
are identified by common base sequences, or boxes. These have come
to be called *homeobox containing genes,* or *Hox genes* for short. A
group of these known as the Hox D cluster is active in limb formation.
They are expressed in a spatial pattern of overlapping domains that
provide five unique combinations of the genes, corresponding more
or less to the positions where the five digits develop, as shown in Figure

159

5.19. The anterior part of the bud is at the top in this diagram, with the tip of the bud toward the left. So this is the left limb bud looking down from the top of an embryo with the region marked "5" corresponding to digit V ("little finger"). The spatial domains of the Hox genes are numbered 1–5, corresponding to genes D9–D13, which are arranged sequentially on a chromosome as shown by the boxes.

It has been shown in the chick hind limb that if one of these genes, such as Hox D11, has an extended domain of expression, covering the region where digit I arises, then this digit is transformed into a digit II-like structure—a typical homeotic transformation. These genes are doing their job in just the same way that the homeotic genes in the floral meristem are: as their spatial patterns change, giving different combinations of genes from the normal, then there is a corresponding change in the structure of the element produced. However, notice another similarity in connection with the *number* of elements produced. We saw that it is not the homeotic genes themselves that decide how many petals or stamens there are. Something else is doing this job, presumably involving other genes. The chick hind limb has only four digits, even though there are five combinations of Hox D genes available. So what the Hox genes are doing is producing *differences* between digits whose number is determined by other processes. However, the maximum number of *distinct* digits specified by the five Hox D gene combinations is five. This has been suggested as the explanation

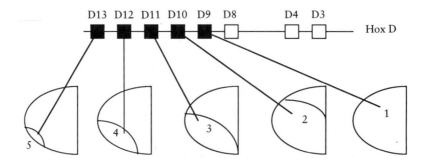

Figure 5.19   *The spatial patterns of the Hox D genes in a chick hind limb, and their relative positions on a chromosome.*

for why tetrapods have no more than five distinct digits. But what about the fossil tetrapods *Ichthyostega* and *Acanthostega*, with seven and eight digits? The proposal is that some of these digits are duplicates, so that they fall into only five distinct categories. This is disputed by some paleontologists, and fossil remains are not easily resolved into the detail required to decide this, so it stands as an interesting possibility only.

The Hox D genes are sometimes described as giving identity to digits. But is it only the Hox genes that do this? Digit III of the chick and digit III of the cat are different—that is, they have different identities, and both are different from digit III in any other species. Clearly, more genes are involved in specifying digit character than just the Hox D cluster. Genes are involved in all aspects of morphogenesis, influencing all stages of the process that starts with the initiation of a field and ends with detailed morphological structure. The differences between digit III of the chick and digit III of the cat are almost certainly due to the action of different genes influencing different levels of the limb field, from initiation and axial specification through the condensation, branching, and segmentation events of cartilaginous aggregate formation, to the lengths and detailed shapes of the various elements. At no stage can we talk of a separate and distinct process of conferring identity on elements.

The Eyes Have It

The last example I present to illustrate my argument about the evolution of generic forms is also one of the most classic.

When Darwin contemplated the design of the vertebrate eye, he had very mixed feelings. On one hand, he was filled with awe at this remarkable result of the evolutionary process; on the other, he found it an enormous challenge to his theory of evolution by natural selection—it gave him a cold shudder, he said. How could random variations ever fortuitously conspire to produce the first functional eye, that initial step required before natural selection could get a grip on it and subsequently refine it into the sophisticated visual systems found throughout the vertebrates and in invertebrates such as gastropods, cephalopods, crustacea, and insects? What is even more extraordinary is that this organ has evolved independently in at least forty different lineages. Eyes seem to pop up all over the evolutionary map, and each time they present the same challenge, provoking the same Darwinian shudder: How could random, independent events ever generate such an inherently improbable, coherently organized process as that required to generate a functional visual system in the first place? What I suggest is that eyes are not improbable at all. The basic processes of animal morphogenesis lead in a perfectly natural way to the fundamental structure of the eye.

Plant form always results from outgrowth, and the full complexity of the organism's shape is visible to the external observer (including the roots, if the plant is grown in liquid or another transparent medium). However, animal embryos can develop complexity by either an outward or an inward deformation of sheets of cells which, lacking cell walls, are flexible and can change shape with or without growth. This also makes it possible for animal cells to actively migrate over surfaces, either as organized sheets, as single cells, or as processes such as axons that grow over surfaces in an organized, directed manner. The result is that animals can develop great internal complexity as well as intricate external patterns. But, as we shall now see, this is achieved by basically the same type of cellular organization as that which operates in plants, though now with the freedom of movement that comes from the absence of a cell wall.

The first major morphogenetic movement of vertebrate embryos is *gastrulation*, the inward movement of cells that is initiated in a particular region of the blastula, the spherical ball of cells that results from cleavage of the fertilized egg (Figure 5.20). A blastopore, or furrow, forms from the invaginating (inward-moving) cells, and deepens as the migrating cells spread over the inner surface of the blastula producing a multilayered structure, the gastrula (Figure 5.21). If the protective vitelline membrane that normally surrounds the embryo is

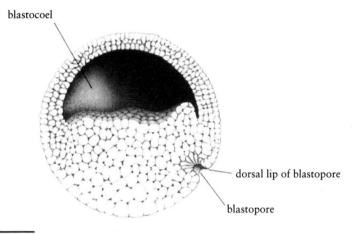

blastocoel

dorsal lip of blastopore

blastopore

Figure 5.20   *Gastrulation in an amphibian embryo.*

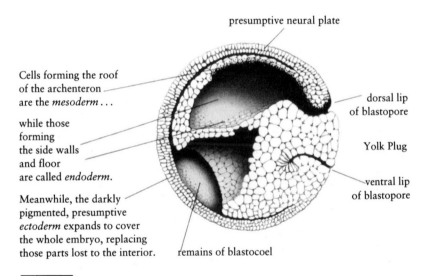

presumptive neural plate

Cells forming the roof
of the archenteron
are the *mesoderm* . . .

while those
forming
the side walls
and floor
are called *endoderm.*

Meanwhile, the darkly
pigmented, presumptive
ectoderm expands to cover
the whole embryo, replacing
those parts lost to the interior.

dorsal lip
of blastopore

Yolk Plug

ventral lip
of blastopore

remains of blastocoel

Figure 5.21   *The structure of an amphibian gastrula.*

removed, and the salt concentration of the medium in which embryos
develop is increased, then instead of invagination followed by migration of cells over the inner surface of the blastula, the blastula buckles
out and cells flow over one another to produce an everted sphere. That
process is called *exogastrulation.* The lack of the interaction between
outer and inner cell sheets, which normally occurs in the multilayered
gastrula and gives rise to further patterns of cell movement and dif-

ferentiation, results in the failure of the exogastrula to develop normally, and a nonfunctional, though coherently differentiated, structure is produced. Under normal conditions, the vitelline membrane and osmotic conditions bias the process strongly in favor of invagination and gastrulation. This shows that the mechanical and ionic conditions of the environment are important in determining which of the options (only two in this case) is followed by the developing organism, just as the ionic composition of the medium is important in determining which pattern of growth (local or global deformations) is followed by *Acetabularia*. Organism and environment together define the developmental dynamic and the morphogenetic trajectory.

The next major morphogenetic movement of vertebrate embryos is effectively a repeat of the first, but now the infolding of the cell sheet takes place along a line and forms a tube from the neural plate as a result of the influence of the inner cell sheet on the outer and the A-P axis that results from gastrulation. The process is called *neurulation,* and the tube formed is the neural tube, from which the nervous system develops. These deformations of cell sheets to produce the processes of gastrulation and neurulation can be simulated by models that treat cells as excitable media of the same kind as that described for *Acetabularia:* the mechanical state of the cytoskeleton alters in interaction with calcium, and cells change their shape, giving rise to propagating waves of cell deformation that result in invagination movements. Figure 5.22 shows such a simulation by Gary Odell and colleagues. Other aspects of morphogenesis in animal embryos based on cytoskeleton–calcium dynamics and its consequences with respect to cell shape change, condensation, and migration have been described by George Oster, Gary Odell, and Jim Murray. Near the anterior end of the neural tube, tissue begins to bulge out laterally on both sides and grow into bulbous structures, the optic lobes (see Figure 5.23). These continue to grow laterally until they reach the surface layer of the embryo, the epidermis. On contact of the optic lobe with the epidermis, the lobe flattens to form the optic vesicle, which then deforms inward in a repeat of the movement that produces the neural

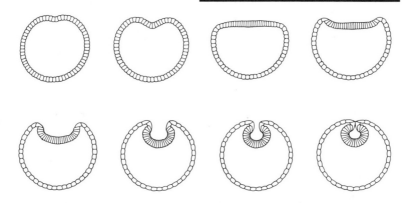

Figure 5.22   *A model simulating neurulation in an amphibian.*

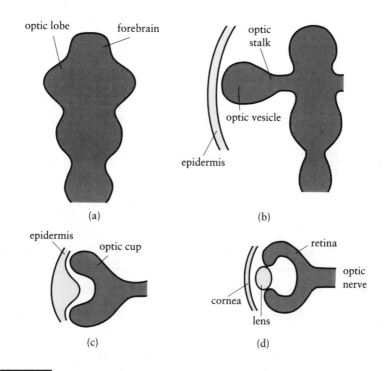

Figure 5.23   *The sequence of shape changes in the developing brain lead-*
ing to an eye: (a) initial appearance of optic lobes, (b) out-
growth toward epidermis, (c) formation of optic cup, (d)
formation of lens.

165

tube, forming the optic cup. As this occurs, epidermal cells respond to contact of the optic vesicle by undergoing a transition from squamous (flat) to columnar, resulting in a thickening and inward buckling of the sheet (as in gastrulation and neurulation), which eventually results in the detachment of the thickened cells to form the lens (just as the neural plate formed the neural tube, but now the geometry is circular rather than cylindrical). The lens becomes transparent, as does the overlying cornea, and the cells of the optic cup differentiate. The inner layer of cells of the retina differentiate into neurons whose axons grow over the inner surface of the retina and down the optic stalk, forming the optic nerve. When they reach the midbrain, they spread out in an organized two-dimensional projection that maps the retina in an ordered manner over the neural tissue that processes visual information (the optic tecta).

The outer layer of cells of the retina differentiate into the pigmented epithelium that responds to photons coming through the transparent, crystalline lens (protein crystals, remember). These epithelial cells store up vitamin A, gift of the plant world, and convert it into the pigment in the rods and the cones that catch photons. The same substance that catches photons for photosynthesis in plants is used to catch them for vision in animals. Such are the bonds of dependence in the living realm. And of course the carbon that makes up the backbone of vitamin A, and all of the materials of life, is a gift of the stars, where carbon and most of the other elements are made. Patterns of relationship run deep in nature.

You might ask why the layer of cells of the retina closest to the lens become the nerve cells that conduct electrical impulses to the brain, since these impulses arise from the layer of cells furthest from the lens that contain the pigment that catches photons. It would seem more sensible to arrange it the other way around, so that photons would not have to pass through a screen of neurons before reaching the pigmented cells. Invertebrates such as the squid and the octopus have highly developed eyes, and they organize the retina with the pigmented cells inside and neurons outside. But invertebrates make their eyes differently—the optic cup arises directly from the surface layer of cells (the ectoderm) by invagination, the familiar buckling of a cell sheet that all animal embryos use to develop internal complexity. They do not make the optic cup from an extension of the brain the way vertebrates do. So in squid and octopus the pigmented epithelium forms from the cells that were on the surface of the embryo, where pigmentation is often found. Vertebrates seem to have a choice of having the pigment in the inside or the outside layer of the retina, since both arise from cells that were originally on the surface. However, it seems that a number of factors are involved in committing the outer layer to being pigmented and developing into rods and cones, among which are mechanical stresses that make the cells go columnar and the blood supply to this layer. And this type of visual system works well enough—extraordinarily well, in fact.

You can now recognize that many of the events that form the vertebrate eye are repeats of the basic movements that we have encountered again and again as the natural-state changes of morphogenetic fields: calcium–cytoskeleton dynamics, localized cell growth and deformation, bucklings of cell sheets, and directed cell movements over surfaces. These do not account by any means for the full detail of the events that give rise to the highly refined visual systems we see operating in vertebrates. However, that was not our goal, which was the more modest one of suggesting how a primitive but functional system for recording visual images could have arisen independently in many different taxa. It is now clear how simple and natural it is for an embryo to generate a structure of the type shown in Figure 5.23(c). With a partially transparent epidermis and excitable cells (neurons) in the optic cup, this already functions as a primitive imaging system, a useful visual organ. This is the first necessary step in the evolution of more sophisticated visual systems, which arise by extensions and refinements of basic morphogenetic movements. The processes involved are robust, high-probability spatial transformations of developing tissues, not highly improbable states that depend on a precise specification of parameter values (a specific genetic program). The latter is described by a fitness landscape with a narrow peak, corresponding to a functional eye, in a large space of possible nonfunctional (low fitness) forms. Such a system is not robust: the fitness peak will tend to melt under random genetic mutation, natural selection being too weak a force to stabilize a genetic program that guides morphogenesis to an improbable functional goal. The alternative is to propose that there is a large range of parameter values in morphogenetic space that can result in a functional visual system; that is, eyes have arisen independently many times in evolution because they are natural, robust results of morphogenetic processes.

That is my case for seeing the major morphological features of plants and animals in terms of the evolution of generic forms. I have presented only a few examples to illustrate the principles. There are many others that come from the studies of a great variety of biologists, and much more work is needed to extend these ideas. The main proposal is that all the main morphological features of organisms—hearts, brains, guts, limbs, eyes, leaves, flowers, roots, trunks, branches, to mention only the obvious ones—are the emergent results of morphogenetic principles. These structures vary within different species, and it is in these small-scale differences that adaptation and natural selection find a role. Even here, however, there may be a significant contribution from generic properties of morphogenesis and pathways of cell differentiation. The balance between intrinsic dynamics of organisms and stabilization by functional demand remains to be determined. Natural selection itself may soon be assimilated into our view of the intrinsic dynamics of complex systems, as we shall see in the next chapter.

chapter

6

New
Directions,
New
Metaphors

Biology and physics use different metaphors to describe the dynamics of change. In biology, the process whereby a new, superior variant becomes established in a population—for example, a bacterium with resistance to a new drug, or a variety of finch whose beak has grown so that it can eat a new range of seeds—is described as an increase in its fitness relative to its competitors. Fitness is the capacity to survive and to leave offspring in a habitat, the organisms that achieve higher levels of fitness being the winners in the struggle for survival. A vivid metaphor for the evolutionary process then follows: organisms are constantly striving to climb up to the higher peaks in the fitness landscape that represents the various possibilities for survival. However, as organisms change, evolving into different varieties and species with adaptations to particular habitats, the evolutionary scenario changes: the fitness landscape itself undergoes modification as evolving species create new opportunities for survival. Trees modify and enrich the soil by dropping their leaves and producing organic compost that retains water, so forest systems such as the Amazon that originally develop on extremely poor soil create conditions for the stunning variety of species that have emerged in this vast ecosystem. The image of organisms striving to climb up local fitness peaks in this evolutionary landscape, which is constantly changing as a result of their own efforts so that they have to keep running just to stay in the same place, fitness-wise, provides a dramatic metaphor of life as continuous struggle to improve merely to survive. This is sometimes seen as progress.

Compare this with the metaphors used in physics and mathematics to describe dynamic processes. A basic example is provided by the image of a marble moving in a bowl. Suppose that the marble is released near the top of the bowl with a push, giving it a horizontal velocity. As it falls toward the center under the action of gravity, it travels around the bowl and gradually settles to the stable point at the bottom where its motion ceases. This point of stability is called an *attractor*, and the bowl defines the basin of attraction for all the trajectories that converge on the stable point. For certain classes of dynamical system, an energy function can be defined that takes a minimum at the stable state. This conveys the idea that natural processes follow paths that decrease energy or some similar function, suggesting that what comes naturally is the path of least effort or action. As a result, we have two quite different metaphors for biological and physical change, the former involving struggle and effort, the latter happening with least effort or action.

Mathematically this does not present serious problems, because descriptions of biological process involving optimization of fitness are formally rather like physical processes that minimize energy: to make them equivalent, the fitness landscape can simply be inverted so that what were peaks become valleys and the fitness functions are converted into types of energy function. However, neither fitness nor energy functions are generic to dynamical systems; that is, they cannot be used to describe the trajectories of dynamical systems in general, being restricted to special classes only. As descriptions of biological process become more sophisticated in terms of the dynamics of complex systems, there inevitably will be a tendency to abandon the use of non-generic fitness functions and to use the language of basins of attraction, attractors, repellers (unstable states), and trajectories. Over the past few decades a host of new concepts and analytical devices have been developed to characterize the properties that have been discovered in nonlinear systems, especially the strange attractors that are associated with chaos. These have provided new tools for modeling biological processes. In the study of evolutionary dynamics, there is now a way of characterizing change in the species composition of ecosystems that replaces fitness functions with dynamically precise, generic properties of attractors and trajectories defined by what are called *Lyapunov exponents*. These specify whether or not a model of a complex ecosystem, say, is stable or whether it can be "invaded" by a new species and move to a new attractor, and whether the new attractor is simple or strange.

The result is a unification of biology, physics, and mathematics that is accelerated by studies in the sciences of complexity and the realization that similar types of dynamic behavior arise from complex systems, irrespective of their material composition and dependent primarily on their relational order—the way the parts interact or are organized. Biology thus becomes more physical and mathematical, putting the insights of genetic, developmental, and evolutionary studies into more precise dynamical terms; at the same time, physics becomes more biological, more evolutionary, with descriptions of the emergence of the four fundamental forces during the earliest stages of the cosmic Big Bang, the growth sequences of stars, and the formation of the elements during stellar evolution. Instead of physics and biology remaining opposites, the former seen as the science of rational order deduced from fixed laws of nature and the latter described (since Darwin) as a historical science, physics is becoming more evolutionary and generative while biology is becoming more exact and rational.

Agents and Causes

The evolutionary metaphor of organisms constantly struggling up the slopes of fitness landscapes, trying to do better than others in order to survive, does capture a particular quality of living process, despite its excessively Calvinist work-ethic image. It represents organisms as agents that actively engage in the process of evolution. On the other hand, the physical notion of doing what comes naturally by seeking the path of least effort, like the marble falling toward the center of the bowl, suggests a different kind of agency, one that tends to harmonize its actions with the environment and go with the flow.

Despite these differences, both the biological and physical metaphors of process have been forced into the same straitjacket of mechanical causation. In neo-Darwinism, organisms actually have no agency, because they do not exist as real entities, reduced as they are to genes and their products, which assemble by local interaction into the structure of the organism. The genes are maintained for their survival properties by natural selection, a cause external to the organism. There is no agency when organisms are purely mechanical consequences of internal and external forces, despite the metaphors of struggle and effort. In physics, a similar causal description of process has replaced any notion of causal agency. The falling marble in the bowl is a paradigmatic example of an entity without any causal power itself, acted upon by an external force (gravity) that causes it to move within the constraints imposed by its environment (the bowl). This picture holds for all particles and fields, so that matter in motion is inert material without intrinsic power, moved only by outside forces.

The difficulties and inconsistencies of this view have been described by the Oxford philosopher of science Rom Harré, who with his colleague E. H. Madden has proposed a return to a more satisfactory theory of causation (see the book *Causal Powers*). In their description, the marble ceases to be something without causal agency and instead is regarded as an entity endowed with "powerful particulars" that give it the capacity to behave in particular ways in particular circumstances. Instead of a world of inert things that get pushed and pulled about by external forces, which is the mechanical view of process, the world becomes one of things with causal powers that are expressed in specific ways in different circumstances. Then "doing what comes naturally" means acting in accordance with the powerful particulars that are expressed in any given set of conditions, and the dynamic description of process in physics involves a particular type of agency. The marble is not just being pushed by an initial force and pulled by gravity. It has its own power that, together with the powers that act in any given circumstance, results in a particular motion.

The view of physical process as dynamic agency and the expression of causal powers in the natural world is similar to the position taken by Paul Davies and John Gribbon in their book *The Matter Myth*. They argue that matter is not the inert stuff that it is often made out to be. Quantum mechanics reveals a world of action that is anything but mechanical. This has been recognized by physicists for decades, particularly since the demonstration by Einstein, Podolsky, and Rosen in the 1930s that the principles of quantum mechanics imply a deep, nonlocal connectedness between particles. But the notion of particles as tiny bits of billiard-ball-like matter that get pushed and pulled about by fields external to themselves is a hard one to put to rest. The relatively recent experimental confirmations by French physicist Alain Aspect of the predictions of Einstein and colleagues has finally forced the realization that mechanical interpretations of basic physical pro-cesses have to be abandoned and replaced by a much more integrated, interconnected view of the dynamics of change at the fundamental level of physical reality. Some kind of influence exists between particles emerging from particular states of union whereby they remain inti-mately connected, so that certain properties such as spin or polar-ization remain correlated between the particles irrespective of how far apart they are in space. The causal agency that is acting here is an expression of particular powers in the particle fields such that the whole system is a unity. Particles are not acted on by forces external to themselves; they are themselves aspects of a single process that is distributed in space and that changes in time according to defined rules—those of quantum mechanics.

Exactly how this quantum mechanical description is to be understood is a subject of considerable debate in physics. The Copenhagen interpretation, originating with Neils Bohr, maintains that the mathematical formulation of quantum mechanical laws cannot be further analyzed in terms of causal principles, while theoretical physicists such as the late David Bohm and his colleague Basil Hiley at Birkbeck College in London argue for a causal interpretation that is based on an explicit representation of nonlocal causality in terms of a field called the *quantum potential*. This describes a unifying principle of quantum action in mathematical terms. Whichever of these views one adopts, or indeed any of the other contenders in the field, the undeniable consequence is that the old mechanical view of causation, of external forces acting on inert particles of matter, is dead. Physics recognizes that natural processes cannot be described in these terms, and that the phenomena we see in nature are expressions of a deeper reality in which apparently separate entities are united in subtle but well-defined ways.

The belief that the natural phenomena we see—the waves on the ocean, the clouds in the sky, or a rainbow—are to be explained in terms of a different, normally unseen level of process where their real causes are to be found is known philosophically as realism. Most scientists are realists in this sense, though there are great differences of opinion with respect to what the "real" level of causes of phenomena actually is. There is, however, no disagreement among physicists that fields are real even if we cannot see them or bottle them or market them. We can see only their effects, such as the orientation of a compass needle in the earth's magnetic field, the fall of an apple in the earth's gravitational field, or the vortex in the bathwater that is an expression of the hydrodynamic field. What Harré and Madden are proposing in their own causal interpretation of natural process belongs within this tradition of realism, but process for them relates to the concepts of causal powers in these fields, or field–particle systems, which embody within them the capacity for particular types of motion in particular circumstances. This is fully in line with the quantum mechanical results relating to nonlocal causality, and is incompatible with any description of matter as inert and lacking in causal agency.

What about biology? Do organisms express the same type of causal agency as inanimate processes, or is there a difference? I think there is a difference, and a fundamental one at that. What is this property of organisms that makes them different from a moving marble, or from a petri dish of Beloussov-Zhabotinsky reagent that generates patterns so much like those produced by organisms themselves? The key is in two related aspects of organismic behavior that can be illustrated by the example of *Acetabularia*, described in chapter 4. If the cap of a mature *Acetabularia* is cut off, a new one is regenerated and the organism is restored to a condition of completion: a part of the original organism, that which is left after the cap is removed, has the capacity to make a whole. This property is also expressed in the life cycle: gametes, which are parts of the original organism, can develop into a new individual. Organisms can regenerate and reproduce. These are expressions of the property of self-completion or individuation that is distinctive to the living state. Humberto Maturana and Francisco Varela, in their book *The Tree of Knowledge,* define this as *autopoesis,* the capacity of active self-maintenance and self-generation that underlies regeneration, reproduction, and healing, which are all ways organisms become coherent wholes. Organisms differ in their regenerative and reproductive capacities. We humans cannot regenerate our limbs the way newts and salamanders can, but we can regenerate skin after wounding, and other tissues such as liver and kidney after damage. But if species are to survive and evolve, the members of that species certainly must have the capacity to reproduce either asexually or sexually. These properties of active self-maintenance, reproduction, and regeneration express a quality of autonomy in organisms: they are due to processes that occur within organisms such that the whole has characteristics distinctive to its particular species.

This autonomy is not to be understood as independence of the environment. Put *Acetabularia* into seawater containing 1 millimole of calcium or a newt into a cold environment ($<5°$ C) and neither will regenerate or reproduce. However, the capacity of these processes for self-completion, and their specificity, come from within the organism itself. The causal powers acting within an organism have a property of closure that defines a specific entity, a dynamic form with identity and agency that can be expressed within a range of environments. In general, nonliving systems do not have these properties. Organisms are endowed with powerful particulars that give them the capacity to regenerate and reproduce their own natures under particular conditions, whereas inanimate systems cannot.

This is an emergent property of life that is not explained by the properties of the molecules out of which organisms are made, for molecules do not have the capacity to make a whole from a part. DNA and RNA can make copies of themselves under particular conditions, but this is a self-copying process, not one in which a more complex whole is generated from a part. This is a principle reason organisms cannot be reduced to their genes or their molecules. The particular type of organization that exists in the dynamic interplay of the molecular parts of an organism, which I have called a morphogenetic or a developmental field, is always engaged in making and remaking itself in life cycles and exploring its potential for generating new wholes. Evolution reveals that these wholes can take many different forms: single cells with complex forms in species such as *Paramecium* and *Acetabularia,* or complex organizations of many differentiated cells in multicellular organisms such as a snapdragon, a newt, a Monkey Puzzle tree, or a human being. And there are many other expressions of this basic property of life, making wholes from parts, as we shall see later. We have now recovered organisms as the irreducible entities that are engaged in the process of generating forms and transforming them by means of their particular qualities of action and agency, or their causal powers. This includes hereditary particulars that give organisms a type of memory, and the intimate relations of dependence and influence between organisms and their environments, as described in Figure 2.4. The life cycle includes genes, environmental influences, and the generative field in a single process that closes on itself and perpetuates its nature generation after generation. Species of organisms are therefore natural kinds, not the historical individuals of Darwinism. The members of a species express a particular nature.

Organisms connect with each other in all imaginable ways, including gene transfer from one organism to another via sexual encounters, direct gene transfer as in microorganisms, or transmission by vectors such as viruses and plasmids that travel from host to host, picking up and delivering genes randomly on the way. This mixing of the gene pool results in an effective search through the potential space of morphogenetic trajectories, an exploration of the possible forms in some of which the living state can be expressed as robust and viable species in suitable habitats. It was suggested in the preceding chapter that the main morphological characteristics of these species, their structural types that allow us to recognize and classify them into categories according to similarities and differences, are the generic forms of morphogenetic processes. These result from robust symmetry-breaking cascades that generate features such as whorls in algae; patterns of leaves and flower elements in plants; and structures such as fish fins, tetrapod limbs, and vertebrate eyes—all the robust, natural results of morphogenesis. An organism of a particular species is an integrated whole with a particular set of characteristics that allows it to function in its environment, pursuing its life cycle. Natural selection is the biological term used to describe the dynamics of the interaction of these life cycles with their environments, which are properly described by Lyapunov exponents, which specify whether the life cycle is stable or unstable—whether the species remains unchanged, expands in distribution, or goes extinct, and whether invasion of an ecosystem by another species can occur. This results in the dynamic image of populations moving toward attractors or away from repellers, rather than moving about in a fitness landscape. It is useful to consider now the significance of the concept of adaptation in this evolutionary dynamic.

Adaptation of Parts or Dynamics of the Whole?

An adaptation is a modification that makes a species better able to survive in a particular habitat. It is generally believed that evolution is essentially a process in which organisms either adapt to changing conditions or die. This tends to give to evolution the property of progress, of species getting better, with more complex adaptations. It is interesting that this notion of adaptation is not used to describe physical processes. Why do we not say that Earth's elliptical orbit around the sun is an adaptation that allows it to perpetuate its dynamic state, to persist in its cyclic motion? Though logically correct, this sounds strange because it ascribes the wrong type of causal agency to planetary motion. But is the agency expressed in evolving organisms best described as one in which there is progressive adaptation to changing environments? That many species possess striking adaptations to their habitats is undeniable: the hummingbird's beak relative to the shape of the flowers on whose nectar it feeds, the legs of the horse that give it such speed on its grassland habitat, the fins of the seal, the coloration and shape of insects that look uncannily like the leaves on which they rest, the bright colors of flowers that attract bees, and so on. But there are just as many characteristics that are not adapted, as described by S. J. Gould and R. C. Lewontin in their paper "The Spandrels of San Marco and the Panglossian Paradigm: A Critique of the Adaptationist Paradigm," in which they discuss the similarities between nonfunctional aspects of architectural design and nonadapted aspects of species morphologies. There are plenty of examples to choose from: whorls of laterals in *Acetabularia* are not adapted characteristics, and neither is your appendix. The fact that light has to pass through the nerve cells of your eye before striking the light-sensitive pigment of the retina is not an adapted characteristic. The digestive system of the giant panda, whose vegetarian diet consists primarily of copious amounts of bamboo shoots whose cellulose it is unable to digest, is not an adapted characteristic. The list, like that of adaptations, could go on for pages. Where does it get us in explaining the dynamics of evolution?

The relevant notion for the analysis of evolving systems is dynamic stability: A necessary (though by no means sufficient) condition for the survival of a species is that its life cycle be dynamically stable in a particular environment. This stability refers to the dynamics of the whole cycle, involving the whole organism as an integrated system that is itself integrated into a greater system, which is its habitat. Focusing on the changes that can occur in the parts of an organism can be informative about the small-scale, or local, aspects of organismic plasticity—the extent to which a bird's beak can get longer or wider, or a butterfly can change its pigment pattern, or whorls can be modified in unicellular green algae. Of course Darwin believed that the large-scale properties of evolution were to be explained in terms of the sum of these small-scale changes. However, as Ernst Mayr, eminent evolutionary biologist, has stated very clearly, there is no evidence for the gradual emergence of any evolutionary novelty by the accumulation of small adaptive modifications. New species seem to arrive on the scene and remain pretty much unchanged during their lifetimes, which can be very long or dramatically brief, an apparent lottery that Stephen Jay Gould describes with his usual virtuosity in *Wonderful Life*. The problem is still the origin of species: adaptation does not explain the major, large-scale features of the evolutionary drama.

Competition versus Cooperation

Another concept that is deeply ingrained in biology is competition. This is often described as the driving force of evolution, pushing organisms willy-nilly up those fitness landscapes to more elevated states if they are going to survive in the struggle with their neighbors for scarce resources. However, there is as much cooperation in biology as there is competition. Mutualism and symbiosis, organisms living together in states of mutual dependency—such as lichens that combine a fungus with an alga in happy harmony, or the bacteria in our guts, which benefit us as well as them—are an equally universal feature of the biological realm. Why not argue that cooperation is the great source of innovation in evolution, as in the enormous step of producing a eukaryotic cell, one with a true nucleus, by the cooperative union of two or three prokaryotes, cells without nuclei? One of these prokaryotes became the nucleus, another became the energy generators (the mitochondria) and, in the case of plant cells, the third type of prokaryote became the chloroplasts. This story of symbiotic union, which is much more complex and subtle than I have described, is one of the most dramatic and far-reaching hypotheses about emergent novelty in the whole of evolution, told in detail and with convincing evidence by its originator, Lynn Margulis, together with Dorion Sagan, in *Microcosmos*. This book also contains an account of the subject of mutualism and symbiosis in the microbial world that points to a new view of interactive dynamics in microbial ecosystems as the foundation of all living systems on this planet. Microbes are the chemical factories of the earth, enriching the soil and the oceans with a vast diversity of products including sulphates, phosphates, and nitrates. They maintain the atmosphere in its nonequilibrium, reactive state by the constant production and utilization of methane, carbon dioxide, oxygen, and other gases; and they are the recycling plants of the planet, converting debris from other organisms into reusable gases, minerals, and chemicals. This microcosmos is the most robust and durable of all ecosystems and will survive most of the disasters the planet may encounter, whether caused by nature or by humans. The realization of the significance of microbial ecosystems for the health of our planet was a major component in the elaboration of the Gaia hypothesis by James Lovelock and Lynn Margulis. Analyzing the dynamic stability of life on earth as an interconnected web of interactions between organisms and their physical environment, they provide precisely the right conceptual structure for a treatment of the evolution of our planet in terms of complex nonlinear dynamic processes, its states of stability, and its points of change.

The immensely complex network of relationships among organisms involves all imaginable patterns of interaction, and there is absolutely no point in focusing on competitive interactions, singling them out as the driving force of evolution. In constructing models of ecosystem dynamics it is certainly important to know whether the effect of one component on another makes a positive, a negative, or a neutral contribution to the rate of change of any particular member of the system. That type of knowledge is essential in the construction of a good model, whether in physics, chemistry, or any other field. At this level of analysis, biology is just like other sciences, and in fact becomes fused with them in the Gaia hypothesis. What this leads to is an understanding of the emergent order of the biosphere—its properties of stability, instability, capacity for regulation and homeostasis, and the effects of nonlinearities on the transitions from one state to another. This can lead to insights into the possible climatic and ecological consequences of global warming, destruction of the rain forests, destruction of species diversity, and so on. Competition has no special status in biological dynamics, where what is important is the pattern of relationships and interactions that exist and how they contribute to the behavior of the system as an integrated whole. The problem of origins requires an understanding of how new levels of order emerge from complex patterns of interaction and what the properties of these emergent structures are in terms of their robustness to perturbation and their capacity for self-maintenance. Then all levels of order and organization are recognized as equally important in understanding the behavior of living systems, and the reductionist insistence on some basic material level of cause and explanation, such as molecules and genes, can be recognized as an unfortunate fashion or prejudice that is actually bad science.

The Evolution of Emergent Order: Life at the Edge of Chaos

The reason competition, selfish genes, struggle, adaptation, climbing peaks in fitness landscapes, doing better, and making progress are so important as metaphors in neo-Darwinism is that they make sense of evolution in terms that are familiar to us from our social experience in this culture. We give our scientific theories meaning by using such metaphors, which, at their deepest level, arise from cultural myths. Darwinian metaphors are grounded in the myth of human sin and redemption, as described in chapter 2, so selfish genes, struggle, progress, good works, and the possibility of altruism for humans are precisely the right images to use in conveying the meaning of evolution from this perspective. This works in a limited way to describe selected aspects of the small-scale changes that go on in organisms and populations. But these small changes do not add up to the emergence of new species, new types of organisms, the difference between *Acetabularia* and *Arabidopsis* or between a shark and a salamander. For this we need a theory of emergent order, one that shows how patterns arise spontaneously from the complex, chaotic dynamics of the living state. Organisms are themselves expressions of this emergent order and agents of higher levels of emergence. The whole spectacle of evolution is this "creative advance into novelty," as Alfred North Whitehead, the philosopher of process, put it. This view is now taking shape in the form of new theories about the dynamic characteristics of the evolutionary process: life exists at the edge of chaos, moving from chaos into order and back again in a perpetual exploration of emergent order. This is powerful, evocative language that has made its way into popular books such as *Complexity: Life at the Edge of Chaos* by Roger Lewin, and *Complexity: The Emerging Science at the Edge of Order and Chaos* by M. Mitchell Waldrop. These describe the exciting ideas and theories being developed at centers for the study of complex dynamic systems such as the Santa Fe Institute in New Mexico.

But what, exactly, does this mean? The conjecture, for which there is extensive evidence but as yet no proof, is the following. For complex nonlinear dynamic systems with rich networks of interacting elements, there is an attractor that lies between one region of chaotic behavior and one that is "frozen" in the ordered regime, with little spontaneous activity. Any such system, be it a developing organism, a brain, an insect colony, or an ecosystem, will tend to settle dynamically at the edge of chaos. If it moves into the chaotic regime, it will come out again of its own accord; if it strays too far into the ordered regime, it will tend to "melt" back into dynamic fluidity where there is rich but labile order.

A number of scientists, many of them connected with the Santa Fe Institute, have contributed to the formulation of this extremely interesting proposition. The first one to formulate the conjecture was Chris Langton, who studied the behavior of cellular automata of the type described by Stephen Wolfram in *Theory and Applications of Cellular Automata*. These automata obey simple rules of interaction like those described in chapter 3, where they were used to model the behavior of ants. They can develop complex patterns of activity. What Langton noticed is that there are some sets of interaction rules that result in ordered patterns in which the system settles to an unchanging state and others that produce extremely complex activities, separated by rules that result in partial ordering. In this partially ordered regime the system is dynamic and changeable, neither chaotic nor "frozen" but richly patterned with activity that extends across the whole space in which the automaton operates. Langton collaborated in this work with Norman Packard, and together they coined the phrase "life at the edge of chaos" to describe the lifelike behavior of their cellular automata and the existence of a transition region that separated the domains of chaos and order. They noticed that in this region all the parts of the system are in dynamic communication with all other parts, so that the potential for information processing in the system is maximal. It is this state of high communication and "emergent" computation that struck Langton and Packard as a condition that provides maximal opportunities for the system to evolve dynamic strategies of survival. The evolutionary scenario here is that the automata change their rules by random permutation, as genes do by mutation and recombination, and that they settle at the edge of chaos as the "best" place to be for maximum adaptability. Langton and Packard then suggested that this state is an attractor for an evolving system, but they did not show that their automata actually went to this region spontaneously. They had to construct an artificial selection, or fitness, function that drove the automata toward that state. However, here is a potentially very far-reaching new idea about the type of internal dynamic properties that characterize the condition toward which a complex system with hereditary variation will spontaneously evolve. The conjecture is that this state is defined by an attractor characterized by maximum dynamic interaction across the system, giving it high computability, to which state the system continuously returns as it explores its changing world.

The automata initially studied by Langton and Packard were too simple to represent a simulation of an evolving system in a changing environment, but in the few years since they first introduced their conjecture, there has been a dramatic flurry of activity that has extended their idea and is producing ever-sharper focus on how to define the attractor for more complex systems. This research is now evolving around a constellation of projects that fall under the name of "Artificial Life" that Langton initiated with the first conference on this theme at the Los Alamos National Laboratory, about twenty miles north of Santa Fe. This is where the atomic bomb was born. The historical irony of the greatest instrument of death and destruction ever devised by humanity and the science of artificial life emerging in the same extraordinarily beautiful region of the planet, a land held sacred by its indigenous people, is not lost on the participants in this new science.

Stuart Kauffman of the Santa Fe Institute is one of the major contributors to the development of this area of study. His recent book, *The Origins of Order: Self-Organization and Selection in Evolution,* documents his work in the field of theoretical biology since his first forays into computer simulations of complex biological systems in the late 1960s. He was one of the first to investigate the dynamic consequences of simple rules governing the activity and interaction of units that he used to represent genes. What distinguishes Kauffman's work is that he is always looking for the robust, typical properties of complex systems, those that emerge without being put into the system in the first place. What could possibly arise from hundreds of genes interacting with no particular order, each gene being randomly assigned one of a limited set of logical functions representing its response to any two other genes also chosen at random? Nothing of interest, you might think. But a quite remarkable result emerged that set Kauffman on an extremely productive research track.

The networks had an immense number of states available to them since each gene could switch on or off, so that a network of 100 genes had 2^{100} states. If we imagine this network wandering through all its states in some sequence, and each state lasts no more than a microsecond, then it would take more than a billion times the age of the universe to get through all the states. This gives you an idea of what it means for a system to be dynamically complex. Cells have thousands of genes, so the number of genetic states available to them is $2^{10,000}$ or more, which is colossal. How can they cope with such overwhelming complexity? The answer proposed by Kauffman is: they get order for free. Here is what happens.

What the networks of 100 genes typically did was to get into a cycle that lasted for an average of 10 states, suggesting a square-root relationship between gene number and mean cycle length. This is an extreme localization of activity. These systems exhibit unexpectedly high degrees of dynamic order. Kauffman then showed that the mean cycle time for networks of up to 1,000 genes, and by extrapolation for much larger ones, has the same square-root relationship to the number of "genes" in a network as the cell division times of cells of different species has to their estimated genome size. He also showed that the typical number of different state cycles of a network varies with the number of "genes" in the same way as the number of different types of cell in an organism varies with its genome size, again involving a square-root relationship. These biological correspondences were very encouraging, especially since the model of genetic networks is so general. The early work then led to studies of networks, with each gene interacting with three instead of two genes, then four, and so on. What emerged was that, as the richness of interactions increased from two to three to four, the dynamics changed from a high degree of localized order to much more prolonged and disordered activity: there is in fact a phase transition from order to disorder very similar to the transition observed by Packard and Langton in their cellular automata as the rules change in certain ways. This begins to look like a general property of complex systems. Furthermore, there is evidence that real genes interact with only a few other genes, so that a case can be made that real genetic networks are close to the transition region.

Kauffman has now extended these ideas in several different directions to simulate a variety of biological processes, particularly ones describing evolutionary change. The guiding light in his work is always to look for properties that are representative of a whole set or ensemble of systems constructed according to simple rules, as in his genetic networks, and to consider what they may be telling us about the generic or typical properties of evolving organisms. His focus is on the types of order that can emerge spontaneously in complex systems and the role of natural selection as an external force that drives the systems into particular states of adaptation. Neo-Darwinism focuses on selection as the primary source of biological order, as we have seen, organisms being essentially survival machines. Kauffman questions this, asking whether there is not a rich source of emergent order that is available for free in complex systems of the type encountered in biology, wherein many units interact in simple ways. The evidence he presents actually goes a significant step further, as he makes clear. Not only is there a lot of order in such systems, but there is no way that it can be avoided as a pervasive feature of living systems. That is to say, much (and perhaps most) of the order that we see in living nature is an expression of properties intrinsic to complex dynamic systems organized by simple rules of interaction among large numbers of elements. This order is generic, and what we see in evolution may be primarily an emergence of states generic to the dynamics of living systems.

So we arrive at the same conclusion as in chapter 5, but by a completely different route. There the question was whether there are generic forms that characterize organismic morphology. The conclusion was that the relational order that arises from the complex pattern of interactions between the constituents of developing organisms, described as developmental fields, results in the emergence of typical or generic forms characteristic of organisms. Kauffman, looking at interactions between genes, draws a similar conclusion about the dynamic order they generate, though his focus is on time rather than space. Together these results point to biology as a domain of emergent order in space and in time that arises from essentially simple organizational principles acting within systems that have properties of causal closure among many interacting units, giving them the distinctive properties of self-maintenance, regeneration, and reproduction described earlier in this chapter. But how do such systems arise in the first place? Is life an improbable accident? Or is it another expression of nature doing what comes naturally, and simply falling into a big attractor that cannot help emerging when conditions are right?

Alchemy and the Origin of Life

Needless to say, Kauffman has also considered this question. The origin of life is often described as a result of the accidental production of extremely improbable polymers such as large DNA or RNA molecules with the capacity to replicate, with the help of proteins. Kauffman approaches the problem differently. He considers networks of simple polymers such as small proteins or RNA to be molecules that have general, nonspecific catalytic activity; that is, they can speed up the rates of chemical reactions, including those involved in making these polymers themselves. He shows that as the diversity of these molecules increases, there inevitably comes a point where all the reactions required to make all the components of the network will be catalyzed by some member of the set itself. This is because, as the polymer size increases, the number of reactions catalyzed increases faster than the number of reactions required to make the polymers. The result is that autocatalytic sets emerge spontaneously, that is, sets of polymers with the capacity to catalyze one another's production so that the whole system functions as a closed set. Such networks with catalytic closure could be the first systems with the properties of self-maintenance and self-replication that were described as essential properties of the living state, and the notion of causal closure takes on concrete form in the structure of an autocatalytic set. These networks also have a rich metabolism of diverse reactions, another basic characteristic of living systems.

These ideas about the origin of life have been greatly extended by others working at the Santa Fe Institute, notably by Walter Fontana. He has developed a logical calculus to explore systematically the emergence of catalytic closure in networks of polymers described as strings that act on other strings, facilitating reactions that produce yet other strings. He calls his theory "AlChemy," short for Algorithmic Chemistry. His results are very significant: self-reproducing *networks* arise naturally and proliferate in the AlChemical world because of their capacity to replicate as connected, causally closed sets. So the analogue of the alchemist's base metals transmuting into gold is simple polymers spontaneously transforming into lifelike systems. Life doesn't need DNA to get started; it needs a rich network of facilitating relationships. This is cooperation, mutual support, and enrichment. Not a bad model for starters.

The concept of life at the edge of chaos may have a much broader significance than its intriguing resonances in biology. A closely related idea was developed by a physicist, Per Bak, to describe a class of physical systems which, like life, are complex, nonlinear, and open to a flow of matter and energy across their boundaries. The generic states of such systems that are far from thermodynamic equilibrium have never been characterized within a general theory, as have states near thermodynamic equilibrium, although Prigogine, who worked on this problem with colleagues in Brussels for many years, developed a clear insight into the emergence of order from chaos in such systems. Bak has described a quite general type of attractor for such systems that he calls "self-organized criticality." This has well-defined characteristics that are encountered in many open systems, in particular a distribution of fluctuations or spontaneous perturbations within such systems that are described as power-law distributions. This means that there is a distinct pattern to the noise or the fluctuations in such systems, revealing a particular type of order. There are significant similarities between these properties and the fluctuations observed by Packard and Langton in their cellular automata when they are at the edge of chaos, with dynamic patterns that also have a power-law distribution and fractal-like properties of self-similarity.

Do Ants Live at the Edge of Chaos?

From these abstractions, let's return to more concrete images to focus our attention on the meaning of organisms as agents of emergence in evolution and look more closely at the relationship between chaos and order in the ant colonies described in chapter 3. Remember that Nigel Franks at the University of Bath, and Blaine Cole in Houston, Texas, observed rhythmic activity patterns in laboratory colonies when the density is sufficiently high, and Cole showed that at low densities the ants behave chaotically. In the computer model that Octavio Mira-montes, Ricard Solé, and I developed, the activity of individual ants is driven by a simple neural network running in chaotic mode so that at low densities they, too, are chaotic. When the density of a colony reaches a critical value in the model, chaos begins to turn into order, and rhythmic patterns emerge over the colony as a whole. As the density is increased further, the rhythm becomes more regular but there is not much change in frequency. Even at complete saturation, when every square of the lattice is occupied by an "ant" so that they cannot actually move to another position, the activity rhythm persists.

We can now ask the question: Where is an actual ant colony on this spectrum of densities? Nigel Franks has made some very interesting observations that bear on this question. He calculated that in a nest there is a nearly constant allocation of about 5 square millimeters per individual so that nest size is proportional to numbers of ants in the colony. Evidently, ants have a sense of density in the colony, since they regulate it at a value around 0.18, with a range of about 0.11 to 0.27.

As ants come and go from the nest, changing the density, rhythmic activity patterns appear and disappear, replaced by disordered activity. This suggests that the colonies are regulating their densities so that they live at the edge of chaos.

The mobile cellular automata model of the ant colony can be used to characterize more precisely the nature of the change from chaotic dynamics to rhythmic activity by using a measure that tells us something about order–disorder transitions. This is the Shannon entropy or information capacity, defined by

$$H = -\sum_{i=1}^{N} p_i \log p_i$$

Here N is the total number of ants in the colony and p_i is the probability of finding i out of the N ants in an active state. The entropy of the system is obtained by taking the sum (Σ) over all these probabilities in the form of the function shown. When this quantity is calculated for colonies of different densities, the result is as shown in Figure 6.1. The entropy has a maximum at a density of 0.24, which is in the middle of the transition between chaos and order. At very low densities, the entropy is small because the dynamic diversity is small, activity tending to be restricted to single ants at any one time even when there are several ants in the colony. With increasing density and higher frequencies of interaction, the activity patterns begin to be more evenly distributed among the possible numbers between 1 and N, and at a density of 0.24 the most even distribution is achieved, making the entropy a maximum. This is the point where there is the greatest variety of dynamic states and all members of the colony are in communication with one another. It is at this point that a colony-wide rhythm of activity emerges. As the density increases above this value there is a tendency for more and more ants to be active as a result of excitations, so the dynamic diversity decreases again and so does the entropy. The colony gets more and more locked into a regular rhythm.

Now let's consider how to interpret these results from the point of view of evolution and the emergence of order from chaos. Are the

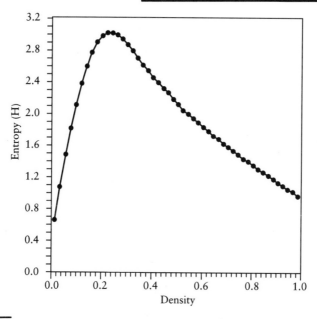

Figure 6.1 *Change in Shannon entropy in a model of an ant colony, showing how it passes through a maximum in the region of the phase transition from chaos to order.*

ants responding to the exigencies of natural selection? Is the activity rhythm with a period of 25 minutes observed in the actual colonies by Cole an optimal value, an adaptation that maximizes their capacity for survival? Or are we dealing with a robust emergent property of a collection of ants, a generic result of their interaction, which they make use of as active agents? There is little doubt that colony rhythms confer advantages such as synchronized feeding and cleaning of the brood, communicating, and getting organized spatially in the ways for which social insects have such a remarkable talent, achieving levels of organization that are miraculous to behold. However, there are also solitary species of ant, so there is nothing obligatory about collective action for survival in the ant world. It is simply one of the possibilities. Rhythmic activity is an emergent property that arises spontaneously in ant colonies, and the ants then make use of this dynamic mode that gives them opportunities for collective action. The rhythms are likely to have very different periods in different species since they depend on the activity characteristics of the individuals, which will differ among species. Some fine tuning is possible. However, if there is an attractor at the edge of chaos that is the most stable and robust state of these dynamic systems, then all ant colonies with rhythmic activity patterns may be expected to have densities that are just on the transition between order and disorder. Then there is no selective force outside the system that "pushes" it up a fitness landscape to a local optimum, one out of many possible states available to the colony. Rather, ants generate emergent order by their interactions, producing a behavioral field with generic attractors that represent states of emergent order, which the ants make use of as dynamic agents. These alternative explanations are open to investigation.

Play and Playlike Behavior

The spectacle of animals at play is a puzzling one from the point of view of natural selection. Imagine two young cheetahs frolicking about in the grass of the savannah, not far from a herd of Thomson's gazelles. They're running, tumbling, feinting, growling—expressing a joy of movement that is totally infectious. But they are taking enormous risks. Lions are constantly on the lookout for young cheetahs, which they ruthlessly destroy. And these two have just scattered the gazelles, one of whom their mother was carefully stalking, anticipating a meal that she badly needed because of the demands of feeding her two rapidly growing dependents. What is the point of this play that apparently reduces the cheetah's chances of survival? They would do much better to carefully copy their mother in stalking, chasing, and catching prey, directing their energies and activities to something useful that increases their chances of survival, which for cheetahs is none too good to begin with. But we see this type of behavior throughout the higher animal kingdom. A troop of monkeys is a familiar example. The amount of energy used by the young in chasing, climbing, leaping, frolicking, and general high jinks is so infectious that you want to join them. Play invites participation. Dolphins cavorting about in the bow wave of a boat or playfully interacting with swimmers provide another familiar example of creatures apparently just expressing their natures and enjoying themselves.

It is in play that we see the richest, most varied, and unpredictable set of motions of which an animal is capable. Compared with most goal-directed behavior, which tends to have strong elements of repetition that give it a somewhat stereotyped, even mechanical, quality, play is extraordinarily fluid. Walking, digging, hunting, eating, courting, and mating all involve well-defined sequences of repetitive movement, whereas play consists of all possible motions unpredictably emerging from one another. Even fish seem to play, or engage in what is called *playlike behavior,* since biologists are rather doubtful whether fish can really play. The complex and fascinating behavior of fish can be illustrated by two males of an Asiatic species of *Barbus* in a tank. These are smallish fish, three to four centimeters in length, making them convenient for laboratory study, but they are highly developed and their behavior is typical of many species of fish. In a tank with a rich and varied range of vegetation, they initially explore the space and each other with complete freedom of movement over the whole tank, independent of where the other fish happens to be. When they encounter one another, the intensity of their activity increases, becoming very rich and varied and running through the full repertoire of which they are capable—turns, darts, reversals, cartwheels, and pirouettes, with occasional fin biting and near collisions, but no focused aggression. When they separate they "cool down." Gradually, they come to an agreement about where they go to rest in the tank, defining the "home base" for their territory. The tank gets partitioned between them so that each fish tends to spend more time in the neighborhood of its base, with a fairly well-defined boundary delimiting the two territories. The original symmetry of movement, each fish visiting all regions of the tank with roughly equal frequency, is broken as the space gets partitioned between them and they visit certain regions more frequently than others.

Another symmetry may also get broken. The initial equivalence of the fish with respect to movement and interaction can change into dominance–subordination, one male having more freedom of movement than the other and tending to initiate bouts of interaction. Even after the two have settled into territories or one has become dominant, they still reenter the state of playlike behavior, which tends to reverse the broken symmetries and reestablish a much more fluid condition of equivalence. Unlike the essentially irreversible symmetry-breaking cascades of morphogenesis that lead to the morphology of the adult form, the broken symmetries that result in orderly, predictable behavior in the fish can be reversed so that a more symmetric pattern of activity is reestablished. Behavior is more fluid than morphogenesis in this respect, though the development of form in both instances can be described in terms of specific symmetry-breaking processes.

The recognition of this equivalent description and its use in the analysis of behavior comes from the highly original work of Koenraad Kortmulder at Leiden University in the Netherlands, who has made a detailed study of fish behavior in these terms. It is his work that gives the insights into playlike behavior in fish and the recognition of their similarities to, and differences from, the dynamics of development. Kortmulder carries the similarities further: he talks about a behavioral field that can be used to describe the dynamics of the space–time patterns of fish movements, and changes in the properties of this field as the fish make the transition from the noninteracting to the interacting mode. As two members of the same species begin to interact they stimulate each other into states of increasing activity and diversity of movement, "heating up" as it were and dissolving previous constraints such as territorial restriction and the dominant–subordinate relationship. This is rather like a phase transition in a physical system with increasing temperature, such as the melting of a solid, the resulting liquid state having more freedom of movement and more dynamic symmetries than the solid. Play, or playlike behavior, is like a high-temperature or excited state, in which the participants have reached a condition of highly unconstrained movement, a state like chaos. As they cool down again, order reemerges from this chaotic condition, usually the same order as before in terms of territory and dominance. Why, after establishing an ordered pattern of relationships in the tank, do the fish reverse it all and go through the process again? When they emerge from another bout of play, the same territories and dominance patterns tend to get reestablished, so the status quo ante is restored. This cycle keeps repeating itself. It is as if the dynamics of interaction between individuals, the pattern of relationship that produces a behavioral field between the pair, is a stronger attractor than those of the separated individuals, so that the relational order of play keeps reappearing. One consequence of this repetition is that out of the high symmetry of play, new patterns of order *can* emerge as the symmetries break again, should these be appropriate. The vegetation in the environment may change so that the territorial distribution may alter; with age, the dominance relationship may transform; and the presence of other fish transforms the behavior patterns. Again we have the suggestion of an attractor, rather like chaos, to which interacting animals of one species keep returning and out of which emerges appropriate order.

The shift of focus in the new biology that is developing out of the sciences of complexity has its focus on the origins of emergent order in complex dynamic systems. This focus resonates with a myth that is deeper than the sin and redemption story, one that our culture shares with all others: the myth of order out of chaos. The dramatic shift of emphasis that accompanies this reorientation in biology merits a new name; I call it a Science of Qualities. The next chapter is an exploration of aspects of this refocused science, which picks up again many of the values that were left behind in the seventeenth century, when the modern age began.

chapter 7

A Science
of Qualities

Is an organism a mechanism? If organisms are mere assemblies of the molecular products of their genes, then there is a good case to be made that, despite their extreme complexity, they are basically molecular machines. This is the molecular perspective, which sees organisms as the result of a genetic program that specifies where, when, and what genes are active in a developing organism and so determines all of its properties. We have seen, however, that molecular composition is not sufficient to specify properties such as the dynamic patterns of excitable media or the forms of organisms that emerge from these patterns. We must understand also the relational order between molecular constituents, the way they are organized in space and how they interact with one another in time, which requires a description in terms of fields and their properties. Fields are fundamental in physics, and it turns out that they are equally fundamental in biology. It is from their properties that the capacity of organisms to make wholes out of parts arises.

One of the clearest distinctions between mechanisms and organisms was given about two hundred years ago by the German philosopher Immanuel Kant. He described a mechanism as a functional unity in which the parts exist for one another in the performance of a particular function. The clock was the paradigmatic machine in his time. Preexisting parts, designed to play specific roles in the clock, are assembled together into a functional unity whose dynamic action serves to keep track of the passage of time. An organism, on the other hand, is a functional *and* a structural unity in which the parts exist for *and by means of* one another in the expression of a particular nature. This means that the parts of an organism—leaves, roots, flowers, limbs, eyes, heart, brain—are not made independently and then assembled, as in a machine, but arise as a result of interactions within the developing organism. We saw in chapter 4 how this happens in *Acetabularia* from an initially uniform field, in which bifurcations occur spontaneously and a cascade of symmetry-breaking processes generates increasingly complex form. In animals, similar processes occur but are accompanied by interactions between cell layers that arise from the systematic foldings and bucklings that generate internal structures and the visible form of the organism. Kant knew nothing about these dynamic processes, but he did correctly describe the emergence of parts in an organism as a result of internal interactions instead of as an assembly of preexisting parts, as in a mechanism or a machine. So organisms are not molecular machines. They are functional and structural unities resulting from a self-organizing, self-generating dynamic.

In the preceding chapter I described reproducing organisms as systems having the property that a part can produce a whole, so that they can regenerate their own natures. The nature of an organism is usually described in terms of the properties of the species to which it belongs, and one of its most distinctive attributes is its particular form.

This form has two aspects: it is spatial, a particular pattern of parts that defines its morphology, such as the shape of an elm or the body of a salamander; and it is temporal, involving particular patterns of activity that define behavior, such as courting in fruit flies or the distinctive flight of a woodpecker. A member of a species is recognized by these two components of its form. These are qualities, the expression of an integrated whole; they are not quantities, a sum of separately existing parts. So we may say that organisms express their natures through the particular qualities of their form in space and in time.

The study of biological form begins to take us in the direction of a science of qualities that is not an alternative to, but complements and extends, the science of quantities. The latter is what emerged from Galileo's studies of moving bodies, in which he focused on what he called *primary qualities* of objects such as their mass, position, and velocity, which could be measured, in contrast to *secondary qualities* such as shape, color, and texture, which are not measurable. The whole of science then became founded on the laws of dynamics (thermo-dynamics, hydrodynamics, quantum electrodynamics), which describe the relationships between measurable quantities such as mass, position, velocity, electric charge, and magnetic force and derived quantities such as energy, entropy, work, and action. Biology has followed this route in describing the quantities of molecular constituents from which organisms are made, and the way these change both during embryonic development as different genes are switched on and off and during adulthood as the physiology of the organism changes from states of rest to activity, from illness to health, or in adaptation to new envi-ronments or learning. These quantitative studies provide very impor-tant information about the dynamic nature of the organism at the molecular level, but they are not sufficient to describe the rhythms and spatial patterns that emerge during the development of an orga-nism and result in the morphology and behavior that identify it as a member of a particular species.

Genes are primary influences in determining which of the possible patterns emerge, but to understand how specific morphologies and behaviors arise in organisms we need to understand the relational order of the living state as described by developmental fields. The emergent qualities that are expressed in biological form are directly linked to the nature of organisms as integrated wholes, which can be studied experimentally and simulated by complex nonlinear models. A striking attribute of fractals and chaotic attractors that has captured the public imagination in recent years is the subtle interplay between chaos and order that we experience as simultaneously beautiful and evocative of living form. This is because organisms themselves are expressions of similarly complex dynamics, from which emerges pattern and order within a coherent whole.

Kant was so struck by the complex and subtle coherence of organisms that he likened the developmental process, the transformation of a simple initial form such as a fertilized egg into the adult form, to the creation of a work of art, which also has an inner coherence expressed in the dynamic unity of its emergent parts. The beauty that we see in organisms he likened to that which comes from the experience of a poem, a painting, or a piece of music. Kant saw this as the world of form, whose enjoyment depends on a free play of the mind, which means not attempting to fix the form in a category but experiencing its coherent wholeness as something of value in itself. "The content here appears in that qualitative perfection which requires no external completion, no ground or goal lying outside itself, and it brooks no such addition. The aesthetic consciousness possesses in itself that form of concrete realisation through which, wholly abandoned to its temporary passivity, it grasps in this very fleeting passivity a factor of purely timeless meaning" (Cassirer, pp. 309–310). An organism or a work of art expresses a nature and a quality that has intrinsic value and meaning, with no purpose other than its own self-expression. Kant described this as "purposiveness without purpose," using the eighteenth-century notion of purposiveness as "individual creation, which displays a unified form in itself and in its structure. . . . A purposive creation has its centre of gravity in itself; one that is goal-oriented has its centre external to itself; the worth of the one resides in its being, that of the other in its results" (Cassirer, p. 312).

The quality of experience elicited is described in terms of play: "The passive stimulation of the emotions is translated into the excitement of their pure play. In the freedom of this play the whole passionate inner excitement of emotion is conserved" (Cassirer, p. 313). Form as a quality cannot exist without quantities that constitute the substantial means of its expression—the molecules of the organism, the stone of the sculpture, the instrument on which the music is played. But quantities without qualities give us a world without beauty, and, as we shall see, without health.

Play, Creativity, and Relationships

Development in human beings involves not only the transformation of the fertilized egg into the form of the newborn infant; the infant also undergoes a prolonged development from which emerges the distinctive behavior and qualities of the adult. This process of child and adolescent development is immensely complex, and I have no intention of attempting to encapsulate it here in a brief compass. I want only to examine two aspects of the process that are particularly relevant to the themes of this book. The first of these is the significance of play in relation to the development of creativity in the child; and the second is the importance of the network of relationships within which the child develops, the field structure of this developmental context, and the ways in which this can give rise to good or ill health.

The relationship between fantasy and reality has preoccupied many a psychoanalyst, because it is where things can go badly wrong for the developing child. Donald Winnicott used his work with disturbed children to develop some remarkable insights into the nature of the transition from what Freud called the pleasure principle to the reality principle in the developing child. Winnicott believed that both Freud and Melanie Klein had attempted to describe this transition in terms of the baby's own development and its connection with external objects, failing to recognize the importance of the mother or the caregiver in this process, which was the focus of Winnicott's own work. A very brief summary of this follows.

The child initially makes no distinction between inner desires and outer reality, experiencing the mother's breast (or the bottle) as a part of its inner reality that magically appears when needed. The infant creates the breast through its desire and imagination and experiences a relationship of love toward it. As the mother becomes less adapted to the infant's needs, it begins to experience the frustration of failure in this magical realm, and the transition to a recognition of the autonomy of outer objects begins. This is assisted by a transitional object, a bit of blanket or a toy or a cloth, that is under the infant's control and is reliably present when needed. The realm between infant and mother becomes a domain of exploration that is not challenged by the question: "Did you conceive of this (transitional object) or was it presented to you from without?" (Winnicott, 1971, p. 12). The transitional object becomes the realm of play, where objects are incorporated into the world of make-believe over which the child has some control. The mother participates initially in this world in which the baby's omnipotence is united with the control of actual objects that are reliable in their constancy, surviving destruction testing in various forms (being thrown on the floor, bitten, lost and found repeatedly). Winnicott describes play as immensely exciting because it is precarious, an interplay of personal psychic reality and the experience of control of actual objects. "This is the precariousness of magic itself, magic that arises in intimacy, in a relationship that is found to be reliable." This develops into the capacity of the child first to play together with the mother, and later to play alone in the security that the mother is there when needed. "It is in playing and only in playing that the individual child or adult is able to be creative and to use the whole personality, and it is only in being creative that the individual discovers the self." Winnicott goes on to suggest that cultural experience itself is located in the potential space between the individual and the environment (including other people) where a creative process occurs that was first manifested in play. Culture, then, is essentially creative play. And the health of an individual depends on the capacity to engage in this process, wherein the deepest creative qualities of human life are expressed. *Homo sapiens* is indeed *Homo ludens,* as described by the Dutch historian Johan Huizinga in his book with that title, *Homo Ludens: A Study of the Play Element in Culture,* in which he argues that "pure play is one of the main bases of civilisation."

However, play is not unique to us; we share it with other mammals, and fish have playlike behavior. "Play is older than culture, for culture, however inadequately defined, always presupposes human society, and animals have not waited for man to teach them their playing," says Huizinga. Play is chaotic and unpredictable, but out of it order keeps emerging. It also appears to share properties with the type of chaos-to-order transitions that are seen in social insects such as ants. Humans may be manifesting a particular form of behavior in creative activity that shares basic dynamic properties with life in general, so that our creativity is essentially similar to the creativity that is the stuff of evolution. Life at the edge of chaos does express with remarkable accuracy our current experience of social and economic disintegration as we move toward some new form of global culture. A biology that has at its center the dynamics of emergent creativity will likely give us more insights into the tortuous turns on this path than one that is based on selfish genes and competition. It is also a good deal more optimistic, since it recognizes that cultural disorder, together with more extensive fields of interaction and communication, can give rise to new levels of coherent, integrative order.

Culture as Fields of Relationships

The developing child emerges from the relationship with the mother or other caregiver into an ever-broadening pattern of relationships with other people who share particular values and so define a cultural context. There has been a strong tendency in recent years to describe the cultural development of an individual in terms of two primary sets of factors: the genetic endowment of the individual; and the influence of the environment, both physical and human. The study of this combination has come to be known as *sociobiology*. This analysis is familiar: it is basically the same as the one that described organisms in terms of their genes and the environment, but that failed to explain how the distinctive forms of organism are generated, and organisms as fundamental entities in biology simply disappeared. So it might be expected that a decomposition of society into the genes that individuals carry and environmental influences (described in this context as "culture") might run into the same trouble, failing to explain the distinctive forms of social structure as a realm of emergent order and resulting in the disappearance of society as a fundamental entity with its own coherent properties. Those become reduced to genes, selfishness, and competition as described in selection of kin groups that have a shared genetic inheritance, social rules explained in terms of genetic fitness, breakdown of community and social structure as autonomous levels of order and substitution of criteria of individual interest, and so on. The genetic contribution to social patterns is certainly important, just as the role of genes in the generation of organismic form is important. Difficulties arise only when the equal importance of relational order, which defines the social field, fails to be recognized so that the appropriate level of organization that is the source of emergent social order is ignored.

Tim Ingold, a social anthropologist at the University of Manchester in England, is one who has recognized this deficiency. His recent work is an exploration of ways to reintroduce the social order, described as a relational field among individuals, as the fundamental generative level within which the developing child experiences the emergence of skills and shared values that make him or her an active and creative participant in a society.

> To remedy the deficiencies of the Neo-Darwinian paradigm, I recommend that we view social life not in statistical terms, as

the outcome of a large number of interactions among discrete individuals, but in topological terms as the unfolding of a total generative field. I have used the term "sociality" to refer to the dynamic properties of this field. Returning to an earlier analogy, these properties stand to genetically and culturally transmitted information as an equation stands to its parameter values. Genetic or cultural variation may be expected to induce evolutionary modulations of the social field, but this is not to say that social forms are in any sense genetically or culturally determined. Culture enables us to account for most of the differences between social forms, but they are linked under transformation by the properties of sociality. Traditional cultural anthropology, however, has fallen into precisely the same error as modern genetics, in supposing that forms are exhausted by their differences. Just like a "gene," the "trait" is a trick concept that converts aspects or qualities of human conduct into substantive parts or components. Thus it is supposed that human individuals, endowed with bundles of cultural traits, have all they need to assemble organised social life. Nothing could be further from the truth. The genesis of social order lies in those domains of consciousness and intersubjectivity that are simply bracketed off by the partition of the human being into genes, culture and behaviour (1990).

So the generative field has a wholeness and coherence from which emerge the qualities of human conduct that constitute the social order. Ingold describes this generative field of human behavior, as expressed in persons, in terms of a relational social order.

I have shown how a theory of persons can be encompassed within a more general theory of organisms, without compromising the role of human agency or denying the essential creativity of social life. This creativity, magnified a thousandfold by the work of consciousness, is but a specific aspect of the universal capacity of organisms to act, in a certain sense, as the originators of their own development. It has been said that, in history, "man makes himself," creating from within the very world in which he is a participant. But man (or woman) is an organism, and organisms generally make themselves, creating as they do a history of life. To arrive at this conception of the organism, however, we need a new biology, or should we say an old one?—for its holistic aspirations are redolent of a pre-Darwinian worldview. It must be a biology that asserts the primacy of processes over events, of relationships over entities, and of development over structure. Organism and person do not then confront one another as specific configurations of matter and mind, "two sorts of independent substances," as Whitehead put it, "each qualified by their appropriate passions." Both are rather embodiments of the total movement of becoming that Whitehead so memorably described as a "creative advance into novelty" (1990).

The child, then, develops into a member of the social order and contributes as a creative participant in its quality of life. But what is this quality that is so valuable and yet so elusive in our culture, something that we feel is slipping away in our headlong rush for increased quantities of consumer goods and technological fixes for problem areas—drugs for better health; fertilizers, pesticides, and herbicides for increased crop yields; more cars for transport problems? We now have the twin crises of deteriorating health and deteriorating environments threatening our quality of life. These crises are intrinsically linked, and both are reflections of the way we think about organisms and their interactions. If humans are to be understood essentially in terms of genes and their products, then illness is to be corrected by manipulating them. The result is drug-based medicine and genetic counseling or engineering. These can be extremely effective in certain circumstances, but medical care based on this approach focuses on illness rather than on health. If interactions among organisms are to be understood primarily in terms of conflict and competition—manifestations of selfish genes—then our relationships with other species are naturally those of a dominant species, and our instinct will be to subdue and control other ones, even to extinction. However, this view of life reflects a perspective on evolution that is deeply influenced by our cultural myth of the fall and redemption. Although this gives meaning to the spectacle of evolution for us in terms of familiar social values—competition, good works, rewards, and progress—it is a parochial and idiosyncratic perspective and describes a biology that is flawed and incomplete. A new biology of the type described by Ingold is one in which relationships are primary in understanding the type of order that can emerge, whether propagating waves in excitable media, cascades of symmetry-breaking processes that give rise to biological form in developing organisms, rhythmic activity in colonies of social insects, or the societies of humans that both engender and depend on the creative activities of persons. The objective is simultaneously a more exact and complete biology, and one that is more relevant to issues related to the qualities that are essential to our existence but have no place in current science. A science of qualities is a science of holistic emergent order that in no sense ignores quantities, but sees them as conditioning rather than as determining aspects of emergent process. It resonates with a different myth, that of creation out of chaos, which is virtually universal and so connects much more directly with the values of other cultures, particularly the indigenous ones that have so much to tell us about the qualities that we are losing— health and balance with the environment.

Health in Hunza

The inhabitants of the Hunza valley in the extreme north of Pakistan, where the valley floor is 7,500 feet above sea level and the Himalayan mountain peaks soar to over 20,000 feet, have been widely recognized as outstanding examples of human health in all its dimensions: biological, social, cultural, and ecological. The valley was until comparatively recently extremely isolated and inaccessible, and the Hunza developed a lifestyle and a relationship with their environment that illustrates in dramatic fashion the type of balance between nature and society that allows the full flowering of human potential. A doctor with the British Army, Robert McCarrison, was assigned in 1903 to service in the north of what was then India. His territory included the Hunza Kingdom, about which he wrote: "My own experience provides an example of a race unsurpassed in perfection of physique and in freedom from disease in general. . . . Amongst these people the span of life is extremely long; and such service as I was able to render them during the seven years I spent in their midst (1903–1910) was confined chiefly to the treatment of accidental lesions, the removal of senile cataract, plastic operations for granular lids or the treatment of maladies wholly unconnected with food supply."

There is very low infant mortality, and families have two or three children at widely spaced intervals so that the period of nursing one child, which is up to three years, is not interrupted by another pregnancy. It is believed that the unborn child will suffer undernourishment if the mother is feeding another child. A visiting doctor, Paul Dudley White, reported in 1964 that after examining a group of men between 90 and 110 years old not one of them showed a single sign of coronary heart disease, high blood pressure, or high cholesterol. They have 20-20 vision and no tooth decay. In a country of 30,000 people there is no vascular, muscular, organic, respiratory, or bone disease! When people die, there is no known cause. How do they achieve this extraordinary level of health?

The Hunza are largely vegetarians, though on feast days they eat some meat, chiefly goat. Their farming methods involve extensive terracing of the valleys, with well-developed irrigation systems that divert the water periodically from the mountain streams and rivers to the terraces, and cultivation of a wide range of cereals and grains, vegetables, and fruit, the apricot being the major product for which they are world renowned. Fruits are dried in the summer and stored with cereals, grains, and root vegetables for use during the winter, which is long and harsh. All organic waste is carefully collected and returned to the soil, which is also fertilized with goat, donkey, cow, pony, and human manure. Goats and cows are not numerous, for they consume too much to be used as a source of food.

The story is told that the Pakistani government warned the Hunza that a huge infestation of insects was expected one year, threatening their crops. They offered pesticides as protection, but the Mir (the ruler) of the Hunza and the elders decided against its use. Instead, a spray of fire ash and water was used, which succeeded in repelling the insects and did no damage to the plants or the soil. However, the Hunza were persuaded once to try fertilizer by a salesman who convinced them that their crop yields would be increased. After two years the farmers discovered that more water was needed to grow fertilized crops, and their cereals and grains dried up too fast over the winter, losing nutrient value. So they returned to their organic methods, and now the use of fertilizer is prohibited.

The Hunza Kingdom is traditionally Muslim, but here too they have their own distinctive customs. By Muslim standards the women are totally liberated, going unveiled, working in the fields in pants, and inheriting property. Whereas alcohol is normally forbidden in Muslim countries, the Hunza make an extremely potent wine from grapes grown in hillside vineyards, which is drunk in copious quantities at festivals. The men are skilled at all construction crafts as well as at their national sport, polo. In Hunza this is a game without rules that involves exquisite riding skills on their famous polo ponies. Any lost teeth among the Hunza usually come from this chaotic and dangerous play, and bones get broken but mend fully in about three weeks. Their physical stamina is also legendary, as illustrated in this quotation from *The Wheel of Health* by G. T. Wrench:

> The illustrious traveller and savant Sir Aurol Stein (1903) was amazed on the morning of June 25th to see a returning messenger who had been sent by the Mir to the political Munshi of Tashkurghan to prepare him for Stein's impending arrival. The messenger had started on the 18th. It was just seven complete days between his start and his return, and in that time he had travelled two hundred and eighty miles on foot, speeding along a track mostly two to four feet wide, sometimes only supported on stakes let into the cliff wall, and twice crossing the Mintaka Pass, which is the height of Mont Blanc.

The messenger was quite fresh and did not consider that what
he had done was unusual (1972, p. 13).

A remote Himalayan habitat is certainly not in itself sufficient to
stimulate this kind of health, vigor, and well-being in people who had
made this type of environment their home. Wrench reports a stark
comparison to neighboring communities.

> Punyal is the first bit of country to the West, going from Gilgit
> up the valley of the Gilgit River to the mountains of Hunza
> on the right. Some sixty miles further westwards is Ghizr, on
> the border of Chitral. The people of Chitral are lazy. They do
> not store food carefully for winter and at the end of winter
> are usually starving. Schomberg's two Hunza attendants
> mocked at the hovels in which the men of Ghizr lived. The
> owners of the hovels replied meekly that they knew their
> houses were squalid and miserable but they could not be trou-
> bled to build new ones. The general assent with which the by-
> standers received this information showed how ingrained
> the laziness was in the Ghizr people. The Hunza and the
> Ghizr are not far apart and both live in similar surroundings
> (1972, p. 12–13).

A similar report concerns another neighboring community.

> The Ishkamanis, whose valley is between that of the Yusinis
> and the Hunza, though living under apparently like conditions
> to their neighbors, were poor, undersized, under-nourished
> creatures. There was plenty of land and water, but the Ish-
> kamanis were too indolent to cultivate it with thoroughness,
> and the possibility of bad harvests was not enough to overcome
> their sloth. They had a number of yaks, but were too lazy to
> load them or ride them or to collect their valuable hairs or
> even to milk them. They had no masons or carpenters or
> craftsmen in their country. Many of them showed signs of
> disease (1972, p. 14–15).

What is described as sloth and laziness in these other groups is probably a direct result of malnutrition and an absence of the social cohesion that is reflected in the communal activities of the Hunza, such as milling grain and extracting oil from apricot kernels, which is done collectively and as a ritual. In a marginal environment such as the high Himalayas, a considerable level of communal order is required to achieve health and stability. Then the community acts as a coherent unit within which health and creativity flourish. Given the similar environments of the Hunza, Ghizr, and Ishkamani, it is extremely unlikely that the dramatic contrasts in social structure and health can be accounted for in terms of genetic differences. It is much more probable that these are to be understood at the level of sociality, the relationships that have been developed and perpetuated among the Hunza and with their environment, from which have arisen the traditions that stabilize their distinctive patterns of emergent social and cultural order. It is the dynamics of the field of sociality itself that needs to be understood if we are to explain particular types of social and cultural order such as that of the Hunza, and it is what we need to pay attention to if we ourselves are to move toward such a condition of health and quality of life. This requires a science of qualities based upon the logic of relationships and emergent order. As Ingold puts it: "Organisms and persons are not the effects of molecular and neuronal causes, of genes and traits, but instances of the unfolding of the total relational field. They are formed from relationships which in their activities they create anew."

The Peckham Experiment

In dramatic contrast to the quality of life and health achieved by the Hunza, an examination of the condition of British youth in the early decades of this century provides some startling revelations. In "The Case for Action," two young doctors, G. Scott Williamson and Innes H. Pearse, stated the following in 1931:

During the war (1914–1918) the number of youths entering the army who had to be classed in the C category of physical and mental fitness gravely disturbed the nation. They were not sick, nor were they healthy. . . . Have conditions not improved since the War? No. The last returns from the Navy, the Army, and Police force show that 90% of recruits offering themselves for service were referred at the first medical examination as not having obtained the necessary standard of fitness. There is no apparent improvement shown since the war in the health of the youth of this country. Let us turn to education and the schools. There the evidence is identical. Official reports state that over 1,000,000 children in this country are too unfit to take advantage of education offered to them by the State and paid for by the taxpayer.

Williamson and Pearse focused on certain social and community influences that they considered to be particularly important in producing this disturbing state of affairs in the nation's youth, in addition to the more obvious problems of diet and nutrition.

The next factor of importance . . . is the change that is being forced upon the home from without by developments that have come about in Western civilisations within the last fifty years, changes that are largely dependent on mechanization. . . . it is necessary to define the sphere implied by the word "home" used in a biological sense. It does not stand for a mere structural unit, the small area enclosed by bricks or concrete that usually comprises the dwelling house of the artisan in towns today. Home represents the functional field the limits of which are determined by the range of the parents' free activity. It includes the family circle to which they belong, the friends and acquaintances they choose, the district they inhabit, the work they do; all, in fact, which might well be included within the designation of "parental circle."

The authors were seeking to define the pattern of relationships in the family and the community that are necessary for individuals to develop normally—to be healthy, among other things. For health, wholeness, is a biological birthright, not a gift bestowed upon lucky people. Kenneth Barlow reflected on these issues in his book *Recognising Health* and examined different ways of looking at the biological foundations of health.

Two perspectives are possible—that associated with Darwin which led to such epigrams as "the struggle for survival," "the survival of the fittest," and, of course, "nature red in tooth and claw." These epigrams all derive from the need of systems which are setting up any living whole, for an intake of organic molecules such as only other living systems (including man) can provide.

The second perspective, instead of concentrating on predation and the competition to which the first perspective leads, considers the way in which all living wholes depend upon the processing of their environment by other life forms which set up their own systems. The soil is the foundation on which plant forms depend, in conjunction with the atmosphere and the sun. The infinitely diverse abilities which varying species develop produce the variety of life but their death is arranged as to allow the cycle to be born again.

The importance of these two perspectives lies in the fact that individual people and indeed nations and societies act in the light of that which they conceive they know. A view that everything depends on the struggle for survival and the devil take the hindmost, makes one set of national policies predominate. When perspective allows for the cooperation and mutual dependence of living structures, policies of a very different kind appear appropriate and will be followed. It will be seen that the two points of view which have previously been distinguished—the medical and the biological—largely correspond to these two perspectives. The struggle for survival naturally leads to consideration of what is wrong, and the medical viewpoint ensues. Ecological cooperation in a habitat leads to consideration of cultivation and what Howard spoke of as the birthright of the crop—health (p. 68).

Howard was a botanist who was talking about plant cultivation. Barlow extended health as a birthright to all organisms, including the human species, seen from an ecological perspective. Barlow's book includes an account of an experiment that Williamson and Pearse carried out to develop their ideas about community conditions for health. In the London borough of Peckham they set up, in 1926, a club that was based on whole-family, not individual, memberships.

> Out of this association of the members all sorts of social activities gradually became established in the club. People began to make friends in a way that we afterwards discovered to have been quite impossible in their homes, where overcrowding is such that any social life whatever is liable to lead to an invasion of the home.
>
> The "Centre," as people came to call it, became in a very real sense a centre of their social life. In this field the doctors moved freely among the people. Here ample opportunity occurred of making observations to supplement those gathered in the consulting room (p. 61).

This initial move in the direction of community health was so successful that an expanded Centre was set up in 1935, the creation of a group of interested doctors and social workers that adopted the name The Pioneer Health Centre. A building was constructed with a swimming pool and a gymnasium, rooms for social activities, and a large kitchen, as well as medical and examination rooms. Again the basis of membership was whole families. What Williamson and Pearse had recognized in their first experiment was that the families who joined had experienced in their normal lives a regulatory environment that severely restricted their freedom to explore social relationships due to fragmentation and the regimentation of work. They sought to create conditions in which people, families, and the community could grow and develop in a way that fostered good health—a developmental rather than a regulatory environment. But of course they did not know exactly what this meant. They were soon to find out. Lucy Crocker, a senior member of the social staff in the Peckham experiment, described their early experiences in the following vivid terms as reported in Innes Pearse's *The Quality of Life:*

> With so much to offer that appeals to children—a swimming bath, a gymnasium, a theatre—we started by assuming that they would be glad to make use of all this in a conventional way in groups with instructors, which was then the only way we knew. So we set about getting timetables of activities.
>
> Walking around the building we contacted the children. A gang chasing up the stairs engaged in some mysterious game would readily stop and talk. Yes; they would like gym classes. Oh yes, they would love to learn to swim, were keen on roller skates. We got their names and ages and we grouped them, girls and boys, seven to nine years, ten to fourteen, fourteen to sixteen, and we fixed times when they said they were free to come. We made sure that the gym and the bath were free and instructors available and we posted up lists for all to see. We thought that that was all that was needed in the way of initial effort, that from then on it was a question of keeping things running. Little did we know!
>
> For this is what happened. Although the whole timetable had been drawn up with the children's wholehearted co-operation, when the week came for classes to start, only the odd child came. The others might appear in the building before the time of the class, or later that day—haring around as usual—and they seemed unable to give any precise reason why they had not turned up. By the end of the first week it was clear they were not coming.
>
> So we had to think again; on the brink of what was to be a long, laborious piece of original research, involving ever fresh thinking in terms of education and of biology before we had it clear. At this point I asked the director if he minded how slowly I moved with the children, if he would give me six months? He said he did not mind how slowly I moved as long as I did move and not stand still. Then started what seemed

like a long period of misery, for what we did was walk round and round and up and down the building, watching the destruction, harassed by every well-wisher saying "Can't you do something to stop the children breaking up everything?" The answer was that of course we could but at the risk of driving them out.

But as we walked we watched, until we began to understand. The children came into the building, not as they went to school, that is because they had to, but because they liked to come and felt that they were entitled to (their parents paid!). They came to do what they liked and to enjoy themselves. It seemed that here in their free time even a swimming bath lacked appeal if its use involved instructions—although the instructors, as people, were well enough liked by them. Also we saw that the times when they could come were often unpredictable and cut into a host of other duties—shopping, homework, music lessons, accompanying brothers and sisters—many more obligations than we had realised. Many things might arise at home or at school to prevent their coming at a given time.

At the same time there were signs that *doing what they wanted to do* was not altogether at variance with what was offered. The swimming bath tempted, the gymnasium with its ropes tempted. But somehow there was a barrier between these streaming children and the surrounding opportunities that they found themselves unable to surmount.

At last we saw the answer. The children must be admitted to the various activities individually, so let them get on with it.

One problem was that the two most attractive places had had to remain out of bounds because they were potentially so dangerous that no-one could take the responsibility of letting unsupervised children into them. If the children could play at curling in the long room where there was nothing but tables, chairs, and ashtrays, what would they not get up to in the lofty gym thick with hanging ropes, rope ladders, poles, vaulting horses? Or the swimming bath—whenever you came into the building there it was, its green-blue water lapping as the eddies from some swimmer reached the edges. But it was dangerously deep at one end; who would dare to let the inexpe-rienced in to drown?

If a child could swim the length of the pool would he not be safe? Then, let any child who can swim a length go into the bath for half an hour at any time he or she likes during the time that children of that age are allowed in the building unaccompanied by their parents. Once arrived at, the solution seems simple and obvious, especially now so many years later; but it must be remembered that at that time educationalists were all thinking in terms of classes and groups. The technical problem was this: children, only some of whom could swim, were coming into the building all at different times. How could we control entry into the swimming bath in this cir-culating hubbub? There was already a check-point into the bath changing-rooms because adults paid to go in, but clearly a member of staff could not accompany each child there to give permission. A child would come tearing up to you in the billiard room, in the laboratory where you might be taking visitors around, out on the concrete where you were on your knees helping a six-year-old to lace up skate boots. We took a piece of paper, scribbled the child's name, "swimming" and our name, and off the child rushed to the man at the gate of the baths—and in about a minute he was "in." He had per-mission from a member of staff who knew him and knew that he could swim.

The permissive piece of paper proved the working clue to the children's use of the Centre. The next four years were spent by us in elaborating the technique and by the children in ever more varied reponse to what we came to see as a challenge to them for the development of skill. About eighteen months after the Centre opened there were at last signs of order; not the quietness due to external discipline but the hum of active children going about their own business (pp. 163–65).

This is the experience of relevant order emerging out of what was experienced initially as children literally in chaos, a process of creative play that is still not incorporated into leisure activity and sport as a primary component of physical "education"—because education still carries with it the connotation of organized rather than self-organizing activity, regulatory rather than developmental order. There is no lack of individual insight into what is required, and educators such as Robin A. Hodgkin have described models of the educational process that are founded on Winnicott's insights and that provide for a "developmental" environment for learning (see *Playing and Exploring: Education through the Discovery of Order*). But rarely are these realized in educational institutions. Another account of the activities at the Peckham Centre, by Sean Creighton, member of the original staff, shows how he was led to implement these principles.

> In my work I was freed from the tiresome restrictions of conventional teaching in schools. I was given a free hand to use the superb facilities in any way that I chose and to encourage the members to do likewise. I was free to move about the building and to observe the tentative explorations of the earliest members when faced with the most luxurious leisure provision that I had ever encountered. The design of the building was a revelation. That alone was, in my opinion, the master key to the liberation I have referred to. Although primarily designed to facilitate the doctors' observations, the open plan and windows through which the main activities could be viewed, gave members the unique opportunity to move about the building, to watch others enjoying themselves, and to be tempted to join in and have a go themselves. There was also the social area through which, over refreshments, they could hear about and share knowledge dispensed by the doctors upstairs—knowledge which could ultimately liberate them from ignorance, inhibitions and fears about their own and their family's physical condition.
>
> My main spheres of activity were in the gym and the swimming pool. I soon discovered that the children were not attracted to formal classes which were too much like school, so quite by chance a game I called shipwreck quickly took over the gym. In this game all the apparatus is spread about and the object is to move from piece to piece without stepping on the floor—which represents the water. This game was so pop-

ular it became an ongoing activity from 4–6 p.m. each day, the children varying in age from approximately 6 to 14 years, joined in and dropped out according to their own convenience and inclination. The gym was usually packed. Toddlers had their own session of a similar activity earlier in the day (quoted in Barlow, 1988, pp. 80–81).

Kenneth Barlow's conclusions about the implications of the Peckham experiment for the achievement of health are as follows:

In the individual there is seen to be growth and development—such development displaying remarkable differentiation. Consideration revealed that in the family there was likewise a potential for growth and development, capable of influencing the members of the family. Over the fifteen years or so during which the lives of member families were under intermittent review it became clear that a further potential resided in the community to which the experiment had given rise.

In each of these three instances, the development of the individual; the development of the family; and the development of the community, there is seen to be a potential. When that potential is realised there is health. Health can accordingly be recognised as the realisation of this biological potential. In each case structures are created and it is the use of those structures, the function to which they are put, which allows of judgment of the excellence or otherwise—the degree of health—which has been achieved. But commonly what is potential is not achieved; what is possible is seen not to have occurred. The result in such circumstances is not health—but ill health. It is ill health which involves doctors. It is the high incidence of ill health which has led public knowledge to confuse health, which when realised stands in no need of the repairs which doctors can offer—with ill health.

The identity, mode of operation and the power, of the processes of health stem from basic biology. Biological processes are able to synthesize themselves by what they take in from their environment in the measure that that environment facilitates the exercise of these abilities. When these matters are recognised it becomes possible both to endeavour to render the environment fit for the exercise of human ability and to promote the development of the abilities which do render the environment fit. All of this requires an endeavour to recognise the constructive potential of our biology; it carries with it the implication that it then becomes possible to promote this. This is an approach which is quite different from the present medical concentration on repairs (pp. 87–88).

This vision of a developmental social environment in which individuals can realize their potential through play and relationships, self-organizing into the appropriate social structures that are continuously changing in response to circumstance, is one we have yet to realize in our culture. We are still caught in the metaphors of conflict, competition, and survival of the fittest in social, educational, and economic affairs, as Barlow noted. There is no denying, however, the momentum of change that is now developing. There is a sense of the breakup of many conventional institutions run along regulatory lines in the areas of health, education, business, economics, government, and politics. This breakup often feels like breakdown, because alternatives do not exist and individuals are left holding the fragments of disrupted lives, with insufficient communal or social cohesion to create appropriate new forms. Chaos in the social and economic order is destructive if it persists too long, and an ideology of individualism promotes prolonged disorder because of the breakup of patterns of relationship that are necessary for new social patterns to emerge. Some structure is required for this, such as the physical and social framework provided by the Centre building in the Peckham experiment and its family memberships. Individuals do not have the resources to create such centers, so there has to be community investment and hence community-based resources—not regulatory, centralized political authority that makes uniform decisions for the country as a whole. Local needs vary.

There are signs that regeneration is actually occurring with the type of holistic, bottom-up approach that worked so well in the Peckham experiment, such as that described in *Radical Urban Solutions* by Dick Atkinson, working on urban renewal in Birmingham. Here housing associations, self-help day centers, and local enterprises are interacting cooperatively with church, school, and community, and there are plans for a neighborhood bank. This is grassroots politics that exposes most current political ideology as irrelevant. It reempowers communities "to take back what's ours by right," as one parent put it. At this point we enter politics and the necessity for a new economic order, which is beyond my scope. Hazel Henderson's *Paradigms in Progress,* Helena Norberg-Hodge's *Ancient Futures,* James Robertson's *Future Wealth* and Edward Goldsmith's *The Way* are among the books that point the way to a different society in which cooperative relationships and quality of life become primary components of economic activity.

The Biological Foundations of Health

The Peckham experiment revealed the importance of a range of different levels of organization in the development of health—the individual, family, community, society, environment, economy—all aspects of the field of relationships and forces that act on the developing individual and either help or hinder the realization of potential in the expression of creative qualities. The language used in the reports of doctors involved in this experiment is filled with biological references and metaphors, because the model of development and health (wholeness) of the organism in a suitable environment is the most tangible image to hold in mind in thinking about the other levels involved. However, the importance of the biological level itself in the developing human fetus and in early infancy in laying the foundations of later health has been dramatically emphasized in studies by a team of researchers in the Medical Research Council Environmental Epidemiology Unit at the University of Southampton, led by D. J. P. Barker (see, for instance, *Fetal and Infant Origins of Adult Disease*). In recent years there has been considerable emphasis on the importance of lifestyle factors such as diet, exercise, and smoking in connection with cardiovascular disease, which is a primary cause of mortality in industrialized nations, particularly among men. The Southampton group noticed a paradox: The lifestyle connection predicts that increasing affluence, resulting in heavier eating and less exercise, should increase the risk of cardiovascular disease. However, the evidence is that these diseases, in England, are more common among poorer people. It is known that regions in the country that currently show high mortality from heart disease used to have high infant mortality. Barker's group wondered if factors that adversely affect infant health might also cause disease in later life; that is, the biological foundations of health may be laid in the developing embryo and the infant. To examine this possibility, the group set up longitudinal studies of individuals by using data more than forty years old that gave relevant information about the fetal development and early infancy of a group of people who had lived all their lives in a certain region.

One example is a group of 5,654 men, born between 1911 and 1930 in six districts of the county of Hertfordshire in England, whose weights in infancy had been recorded. This is a properous region of the country, and rates of ischemic heart disease (due to decreased blood supply to heart tissue caused by impaired circulation, discussed in chapter 3) are 18 percent below the national average. There was a strong correlation between the weight of the children at one year and probability of death from ischemic heart disease. Among those whose weights were 18 pounds or less, mortality rates were almost three times greater than among those who attained 27 pounds or more by the end of their first year. There was also a strong correlation between low weight and deaths from all causes, hence life expectancy. The conclusion is that "an environment which produces poor fetal and infant growth is followed by an adult environment that determines high risk of ischemic heart disease." There is no correlation between early growth and death from lung cancer, so smoking does not appear to be a factor that contributes significantly to death from ischemic heart disease in someone who had poor infant growth; that is, poor nutrition during early years puts an individual at risk of heart failure in later life whether or not the individual smokes. So "measures that promote prenatal and postnatal growth may reduce deaths from ischemic heart disease." Because birth weight is strongly influenced by maternal height, which is itself largely determined by growth in early childhood, promotion of growth in infant girls "may lead to improved prenatal growth in their babies and may further reduce deaths from ischemic heart disease."

Another significant result of the Southampton group studies points to the importance of the condition of the developing fetus as the foundation of later health. The adult condition examined in this case was hypertension (high blood pressure due to "hardening of the arteries" and obstruction of blood flow by fatty deposits in arteries and veins). The data used was weight of both babies and their placentas, which had been recorded at a hospital in Preston, Lancashire, for 449 births between 1935 and 1943. The adult men and women were still living in the district and were forty-six to fifty-four years of age when the study was conducted in 1989. They were asked to participate by allowing a field-worker to measure their height, weight, blood pressure, and pulse rate. A surprising result emerged. There was a strong correlation between the ratio of placental to baby weight at birth and hypertension. That is, the highest blood pressures occurred in men and women who had been small babies with large placentas. It was also found that higher body mass index (weight to height ratio) in the adult (tendency to being overweight) and alcohol consumption also correlated with hypertension, as expected from other studies. However, the connection between placental-to-birth-weight ratio and blood pressure (hypertension) was independent of these lifestyle factors (diet and drinking), and it was stronger. The same was true of salt intake, which correlated with hypertension, but much less strongly than the placental-to-birth-weight ratio, and independently of it. Therefore, although lifestyle is certainly a factor in hypertension, factors affecting fetal and infant growth are considerably more important. How is this to be explained?

Animal studies have shown that, in response to hypoxia (reduced oxygen supply to the fetus), more blood goes to the head and less to the body. The Preston data showed that greater placental weight for any birth weight was associated with a decrease in the ratio of length of body to circumference of head; that is, large heads relative to body size correlate with large placentas. The placenta also tends to grow larger if oxygen supply or nutrition to the fetus is reduced, to compensate for these deficiencies. So a large placental-to-birth-weight ratio suggests a reduced oxygen supply, nutrient supply, or both to the fetus, with a consequent redistribution of blood flow that favors the head over the body. As a result, the arteries within the developing body experience reduced blood pressure, and they become thin, muscular, and inelastic, compared with developing arteries that experience higher blood pressures, which are thicker, have a wider lumen (channel), and are more elastic. Therefore, reduced blood supply to the developing fetal body could result in irreversible changes to the body arteries, leading to high blood pressure in the adult. There is no evidence that the data is explained by a genetic mechanism that determines both the blood pressure of the child and the growth of the placenta.

The conclusion is that the intrauterine environment of the developing human has a dominant effect on blood pressure in the adult. "Reducing blood pressure in a population may partly depend on improving the environment of girls and women, including improving their nutrition." The Hunza knew this intuitively, making sure that no nursing mother became pregnant so that the fetus had to compete for nutrition with a sibling. Their social customs as well as lifestyle eliminated hypertension and cardiovascular disease. The message once again is the obvious one that the health of the nation will improve if resources are distributed among members so that there is no serious deprivation. Responsibility for health is not in the hands of individuals alone. The community and society must ensure adequate nutrition of its members and an absence of environmental stresses such as air and water pollution that damage fetuses in irreversible ways. Innumerable factors of social and economic stress, such as those associated with unemployment, homelessness, and personal isolation, have increased dramatically in the past decade, acting on pregnant mothers and parents, and through them on fetuses and children. These are in considerable measure consequences of a focus on growth in consumer goods production and mass consumption as the solution to social and economic problems, rather than on balanced investment in the basic infrastructure of society—homes, public facilities and transport, hygiene, education, health services, and leisure centers as well as innovative technology and consumer goods. The lack of balance in our society is a reflection of a predominant ethos of quantity rather than quality that is encouraged by the reductionist tendencies of current science and its technological applications. If these influences were confined to our own culture, they would at least have a limited domain of impact and we could seek solutions to our own problems. However, the shrinking of the globe puts all cultures at risk as we export our problems to other nations and peoples in the name of solutions. Nowhere is this more evident than in the methods we propose for increasing food production in the third world.

Monocultures of the Mind

The extraordinary achievements of the Hunza and many other indigenous cultures in maintaining a high quality of life through a balanced, sustainable relationship with their environment based on appropriate scientific and cultural practices have all been put at risk, and many already destroyed, by the impact of our own cultural values on these delicately balanced societies. Interaction between cultures produces grist for the mill of human evolution, whose products give rise to the elements of new, transformed social and political structures that facilitate novel expressions of human creativity. There is constant exchange and barter across cultural frontiers, but many indigenous cultures have managed to develop, in direct conflict with our own restless impatience to embrace novelty, wise methods of testing the efficacy of innovations. A new variety of maize or a new crop intro-duced into traditional farming practice would be studied closely for its effects on the rest of the agricultural system over a considerable period of time, often as long as seven generations, before a final de-cision was made about its value in the cultural context. The sensitivity of ecosystems to possible damage is well recognized, and careful meas-ures are taken to ensure that any change would be both beneficial *and* sustainable. Sometimes innovations are so obviously destructive that they can be excluded quickly; witness the speed of the Hunza decision to ban the use of chemical fertilizers from their country because of the increased irrigation required and the reduced nutrient value of stored cereals. This is holistic agriculture in practice.

Compare this to the experience of many farming communities in India during the so-called green revolution of the 1960s and 1970s. This was the application of "scientific" principles to crop production: farmers used genetic varieties that yielded increased amounts of grain (high yielding varieties, or HYVs) and that were particularly responsive to and dependent on chemical fertilizers. These varieties were all from hybrid seeds, crosses of pure-breeding genetic strains, which were sold to the farmers along with chemical fertilizers, pesticides, and herbicides (because these hybrids did not have the natural resistance of indigenous varieties to pests and disease). The increased costs to the farmer were partially offset by government loans, which were then to be repaid out of profits from the increased crop. This worked, often dramatically well, in the short term: food production soared in India so that it became a net exporter of wheat and rice instead of a net importer. However, in the long term the ecological and social costs have been disastrous. As discovered by the Hunza, fertilized crops are thirsty, particularly the HYVs, so that increased irrigation was required. This resulted in the construction of large dams, displacing thousands of people from their homes and causing erosion of flooded areas. Many of these dams have a limited lifetime of twenty to thirty years due to silting up from erosion, as a result of which they become nonfunctional and have to be abandoned. The chemical fertilizers, pesticides, and herbicides cause pollution of drinking water and deterioration of the soil due to an absence of organic materials that preserve soil texture and maintain microorganisms required for ecological balance. The HYVs are short-stemmed, with large heads of grain, so that increased food is obtained at the cost of decreased straw and fodder for farm animals. And because the HYVs are hybrids, their seed does not breed true. The farmers are not able to use part of one year's crop to sow the next, and must obtain the seed directly from the supplier. The overall result is a pernicious cycle of increasing ecological damage, dependency of the farmer, dislocation of normal integrated farming practices, and debt. Farmers become enslaved to "scientific" methods of production that are intrinsically unsustainable, and new techno-logical "fixes" are required to sort out new problems. The consequence is continuous disruption, disturbance of balanced ecosystems, and despair on the part of indebted farmers—many have even committed suicide, using the pesticides to do so.

One would think that this type of experience would be sufficient to allow wiser counsels to be restored, so that any further innovations would be examined and tested with more caution and with a balanced assessment of long-term benefits and disadvantages. However, the ac-knowledged deficiencies of the green revolution are now to be fixed by new technological tricks. Since genes and environmental influences are the only causal factors recognized in modern biology, these are the only variables that can be manipulated in getting organisms to perform better according to our criteria of utility. We now have the technology to alter the genetic composition of organisms directly, by taking genes out of one species and inserting them into another, so genetically redesigned varieties can be produced. These are *transgenic* varieties. There are species of plant that are naturally resistant to herbicides because they have genes that produce enzymes that destroy the herbicides. These genes can be identified, isolated, and transferred to other species. A number of transnational companies that were heav-ily involved in producing fertilizers, herbicides, and pesticides for the green revolution are now busily engaged in the design of crop plants with tolerance to their own brands of herbicide and pesticide. Ciba-Geigy has designed soybeans to be resistant to their Atrazine herbicide, while Dupont and Monsanto are developing crop plants with tolerance to their herbicides. These chemicals are lethal to most herbaceous plants and so cannot be applied directly to ordinary crops. Genetically altered varieties, however, will be able to prosper while "weeds" wither away in herbicide- and pesticide-treated fields. The problems associated with this strategy are so obvious that it is difficult to believe it is seriously entertained after the experience of the green revolution. The herbicide- and pesticide-resistant genes introduced into crop plants are unlikely to stay in these species, and will be transferred by viruses, bacteria, and fungi to other species so that "weeds" will themselves become tolerant, necessitating higher doses of herbicides to destroy them. The result will be more toxicity in the soil and in drinking water, greater ecological destruction, and ill health.

It is claimed that genetically engineered agriculture will be chemical-free and ecologically safe. However, continued use of herbicides and pesticides is in direct contradiction to this. The use of genetically engineered varieties opposes the basic ecological principle of species diversity, which is also the foundation of traditional agriculture. The very concept of "weed" is not traditional, because all plants have their uses. According to a "scientific" report, a particular farm in Mexico had over 214 "weeds," but the farmer had a specific use for every plant. Even the recognized food crops of traditional agriculture are extremely diverse; a variety of different strains of species are used because they have different levels of resistance to desiccation and to pests so that, as conditions fluctuate from year to year, there will always be an adequate harvest. Instead of relying on one designed variety that requires uniform conditions for a successful crop, resulting in catastrophic failure and famine when those conditions are not satisfied or when a new variety of pest makes its way onto the scene, traditional methods use diversity to achieve stability.

The contrast between the monoculture mentality and the ecosystem diversity approach is described in great detail in the book *Monocul-*

tures of the Mind by Vandana Shiva, whose title I have used for this section. The book is a powerful, coherent indictment of Western "scientific" agricultural and forestry practices, which are based on the idea of single crops that maximize a single product, such as wheat or rice seed, wood pulp, or timber. This monoculture mentality arises directly out of a reductionist science of quantities that looks at species in terms of specific traits that can be amplified to give high yields of particular products: milk from cows, seed from grain, wood pulp and timber from eucalyptus (Figure 7.1). Most of the connections between these products are broken. Traditional use of natural products integrates a variety of them into a whole, balanced, and sustainable system

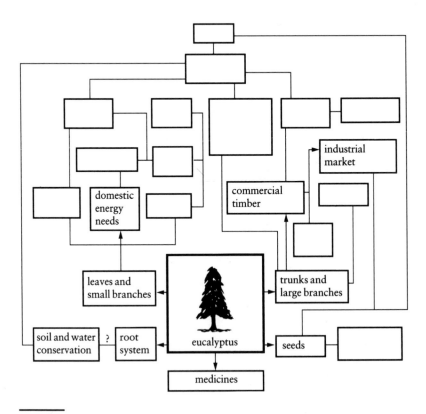

Figure 7.1   *Use of monocultured species introduced for very specific purposes and serving limited interests.* (Source: *Shiva, p. 38.*)

that is intrinsically robust because of the many components of inter-action (Figure 7.2).

What motivates the new genetic engineering revolution is not sus-tainable agriculture but the profits to be made through intellectual property rights on genetically designed strains of plants and animals, continued sales of herbicides and pesticides, and the progressive con-trol over farmers by indebtedness that has been a primary result of the green revolution. The diversity of natural strains and varieties cultivated and preserved by traditional agricultural methods have been turned into banks of "germ plasm" (a legacy of Weismann's division of the organism into a controlling genetic part, and a controlled body part) made freely available to the transnationals who use them as a

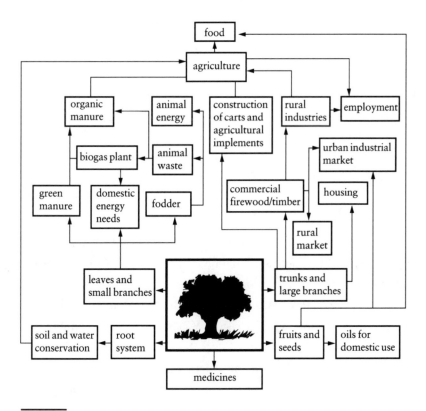

Figure 7.2   *Use of trees in traditional agriculture, serving a great variety of needs.* (Source: *Shiva, p. 37.)*

genetic resource to produce varieties that they then control by patent rights. Even though these varieties breed true so that seed from one year's crop can be used to sow the next, unlike the hybrid varieties of the green revolution, the farmers are now obliged to pay royalties so that their debt continues. Germ plasm thus comes to control not only organisms but also farmers, all in the name of "development."

One of the fundamental misunderstandings underlying many of the so-called development programs is the notion that traditional farming practices based on biodiversity have low productivity. However, when all the multiple yields of diverse crops, the values and outputs of biological systems, are taken fully into account, agricultural practices based on diversity are more productive, produce higher nutrient value in the food than monoculture farming, *and are sustainable.* This diversity includes the use of reeds and grasses for weaving baskets and mats, which is eliminated by the use of herbicides that destroy these plants of the natural ecosystem. And there are countless other instances that contrast the monoculture mentality of "scientific" Western agriculture with the robust, supportive, diverse, and sustainable qualities of traditional agroforestry. Systematic comparative studies being carried out by many research centers are accumulating clear evidence of the superiority of diversity-based agricultural methods, and resistance is growing to the further introduction of inadequately tested innovations.

There is no rejection of novelty in itself. Plenty of technological developments are making their way into village life in India and elsewhere that are perfectly compatible with an integrated lifestyle that seeks quality of life and sustainability. Efficient cookstoves and brick-making equipment, novel looms and weaving techniques, and small dams that save labor and are under local control are all welcomed. These conform to Gandhi's vision of a village-based economy and local empowerment, in contrast to the big fixes that result in big disasters, such as large dam construction and the large farms encouraged by the green revolution that displace local farmers and destroy village life. Communities are the source of health and a high quality of life. The people affected need to participate in adequate testing of any innovation before a decision is made on the advantages or dis-advantages, as in the Hunza experiment with fertilizers. Traditional cultures have useful guidelines for adequate assessment of change that might affect sustainable use of their ecosystems, often extending to several generations. As described by Vandana Shiva in another book, *Staying Alive: Women, Ecology, and Survival in India,* these communities embody the perennial principles that are expressed in Indian cosmology in terms of the dialectical play of opposites: creation and destruction, cohesion and disintegration. Primordial power is called nature (*Prakriti*), which is an expression of both Shakti, the feminine creative principle of the cosmos, and the masculine principle of form, Purusha. All the forms of nature are the children of Prakriti, who is also called Lalitha, the Player, because *lila,* or play as free spontaneous activity, is her deepest nature. The will-to-become-many is her creative impulse, and through this impulse she creates the diversity of the forms of nature. The creative force and the created world are not separate and distinct, nor is the created world uniform, static, and fragmented. It is diverse, dynamic, and interrelated. This is a vision of order emerging from and returning to chaos in a creative play of forms. Each of these forms of life, each natural species, has intrinsic value and meaning in relation to the whole tapestry of life so that there is a sense of the sacred in this living dance. A science of qualities recognizes the value of the emergent properties embodied in the natural forms of life.

Qualities, Values, and Laws

Organisms are intentional agents, engaged in expressing their natures, which have a "qualitative perfection which requires no external completion," as Kant put it. Their worth resides in their existence rather than in their results. This is what we see when we understand organisms as part of the complex matrix of relations from which the forms and patterns of living order emerge, as described in chapters 3 through 6. The recognition of species as natural kinds, expressing the generic forms of living nature that arise by creative emergence from a dynamic

231

process located at the edge of chaos, carries with it implications about our relationships with the species making up the intricate web of mutual dependence that makes our life on this planet possible. These are now practical issues of life and death that are intimately connected with our attitude toward nature. If organisms are seen as mechanisms, they will be treated as such, and as such we will treat each other. The very concept of health, of wholeness, disappears, just as organisms do from modern biology. A biology of parts becomes a medicine of spare parts, and organisms become aggregates of genetic and molecular bits with which we can tinker as we please, seeing their worth entirely in terms of their results, not in their beings. This is the path of ecological and social destruction.

There is a new biology in the making, however, and with it a new vision of our relationships with organisms and with nature in general. This has been expressed by Gunther Altner in the following way: "The prime obligation of human beings towards their fellow creatures does not derive from the existence of self-awareness, sensitivity to pain or any special human achievement, but from the knowledge of the goodness of all creation, which communicates itself through the process of creation. In short, nature imposes values because it is creation." This perspective is an integral part of our cultural heritage that is considerably deeper, and more fundamental, than the sin/redemption metaphors that animate Darwinian biology and much of our social and economic life. It connects us to roots that lie in original blessing (creativity) rather than in original sin, as Matthew Fox has made eloquently clear in his book *Original Blessing*. It further connects us with equivalent metaphors that exist in all other cultures, such as those of creativity and play in the form of Lalitha, the Player, who creates the diversity of the forms of nature. So this offers a way out of a manipulative, quantitative biology based on a parochial cultural perspective, and toward a science of qualities that not only is more complete and exact than current biological theory, but also emphasizes the wholeness, health, and quality of life that emerge from a deep respect for other beings and their rights to full expression of their natures. This is not an idle utopian vision, but the foundation of a sustainable relationship with the natural world. Gunther Altner describes how this perspective leads us to consider a number of very acute problems that are, in fact, currently the focus of intense discussion and debate. Among these he identifies the following.

1. Human history and the history of nature are part of a comprehensive process. The rapid dynamic of human history is threatening to tear apart the indispensable ties which bind us to the history of nature, which runs more slowly. For this reason, moratoria (pauses for thought) are indispensable, so that we can examine the unforeseeable consequences of science, technology and progress. For such moratoria to be regulated there is need for a democratically legitimated process of institution and control with the participation of the critical public.

2. One special problem is posed by the possibilities of intervention provided by modern biotechnology, especially gene technology and the biology of procreation. If living beings have a right to life and to procreation in line with their species, interference with heredity and the reprogramming which that produces is extremely problematical.

3. The rights of nature that are to be called for make it quite essential that the whole sphere of the use of organisms (animals and plants) should be subjected to a critical survey. Here we have on the one hand the question of a proper preservation and procreation of species. And on the other hand there must be a discussion of the function of animals as a source of food and as potential material for medical and consumer experiments (e.g., in connection with cosmetics).

4. There must be an end to the undervaluation of nature in theoretical and practical calculations which regard it

as a resource that is available more or less freely. The rights of nature must be shaped in such a way that nature is taken seriously as a "third partner" in business alongside labor and capital.

5. A limit is put to human existence by the biosphere as the extreme framework for human action. However, this limit remains variable depending on what demands are made on it. . . . Rights can be given successfully to nature only if this intention permeates all realms of law and levels of structure within the biosphere, from local government through state constitutional law to international law.

These considerations lead directly to problems of legislation addressing the rights of future generations and of nature in general. A very useful attempt to formulate these has been presented in the Berne Draft Resolution, among which are listed the following points.

1. Future generations have a right to life.
2. Future generations have a right not to be manipulated, i.e., to have a heredity which has not been artificially changed by human beings.
3. Future generations have a right to a varied world of plants and animals, and thus to a life in a rich nature with the preservation of an abundance of genetic resources.
4. Future generations have a right to clean air, to an intact ozone layer, and to an adequate heat exchange between the earth and the atmosphere.

Included in this draft resolution are the following "rights of nature."

1. Nature—animate or inanimate—has a right to existence, i.e., to preservation and development.

2. Nature has a right to the protection of its ecosystems, and of the network of species and populations.
3. Animate nature has a right to the preservation and development of its genetic inheritance.
4. Living beings have a right to life in accordance with their species, including procreation, in the ecosystems appropriate to them.
5. Interventions in nature need to be justified.

These proposals actually give to nature the status of a subject in law. They lead to the same rights for nature as for humans. The radical program of legal and constitutional action that follows from them is a way of healing the deep wounds that have arisen from our view of nature as something other, something outer, something objective that does not share our values. In current biological science we try to turn organisms into objects. However, it is impossible to deny the expression of pain in a cut or a maimed animal, and so we conduct surgery using an anesthetic, as we do for humans. Pain is a subjective experience, not an objective, quantifiable state. The same is true of the boredom expressed by a bear pacing mechanically to and fro in its compound in a zoo, or a tethered animal that repeats the same movement of frustrated confinement, chewing and pulling on its chain in a stereotyped manner that damages itself. Organisms deprived of a sufficiently rich environment in which to express their natures suffer permanently impaired behavior—a polar bear continued to pace an oval circuit the same size as its old circus wagon, even when put in an interesting environment. Deprivation produces disorder.

Frustration, boredom, and pain are all subjective experiences, but they are no less real for that. A science of subjective states is a real possibility, as Françoise Wemelsfelder has convincingly argued in her book *Animal Boredom*. She quotes from the paper "What Is It Like to Be a Bat?" by philosopher Thomas Nagel: "An organism has conscious mental states if and only if there is something that it is like to *be* that organism—something it is like *for* the organism. I want to know what it is like for a *bat* to be a bat." Wemelsfelder goes on to comment: "To be a subject according to Nagel, is to have a particular, personal point of view on the world, to know the world from 'inside.' An objective view, on the other hand, seeks to detach itself from such a personal point of view, in order to acquire a more general, 'external' perspective on the world. Thus, Nagel conceives subjectivity and objectivity as epistemological notions, that is, as *explanatory perspectives* on the world. . . . Philosophers also use the terms 'first-person' perspective and 'third-person' perspective to denote the subjective and objective perspectives respectively."

Wemelsfelder is working with farm animals on a project that seeks to demonstrate what it means to have a science of subjective states. Again, this is a radical move away from an exclusively objective science, acknowledging the reality of "first-person" perspectives not only for humans but for animals in general. This is also part of the process of healing our relationship with nature, because recognizing qualities and intrinsic values in other beings requires that we acknowledge their capacity to experience life in terms that are similar to, though not identical with, our own subjectivity. Each species, in fact, has its unique relationship to the world, its own experience of what it is like to be itself as an intentional agent engaged in expressing its nature in the context of a particular environment.

Of course, the further implication of this position is not only the reality, but the primacy, of first-person, subjective experiences. This leads toward a science of consciousness. It is another of those curious paradoxes that a large number of scientists who work in the area of artificial intelligence, and in the cognitive sciences generally, deny that consciousness has any fundamental reality and say it is basically an epiphenomenon of brain activity—the electrical and molecular processes that go on in brain cells. This is just like the denial on the part of many biologists that organisms have any fundamental reality that cannot be explained by genes and molecular activities. However, the recognition that organisms have intentional agency that comes from the self-completing action of immanent causal process not only gives organisms a reality that is not reducible to their parts, but also creates a space for subjective experience—what it is like to be a bat or any other species of organism. For us it is the experience of being human, and the awareness of the condition of consciousness. A science of qualities is necessarily a first-person science that recognizes values as shared experiences, as states of participative awareness that link us to other organisms with bonds of sympathy, mutual recognition, and respect.

references

Altner, G. *The Community of Creation as a Community in Law*, pp. 1–10 in *Naturvergessenheit: Grundlagen einer umfassenden Bioethik*. Darmstadt: Wissenschaftliche Burchgesellschaft, 1991.

Atkinson, D. *Radical Urban Solutions*. London: Cassell, 1994.

Barker, D. J. P. *Fetal and Infant Origins of Adult Disease*. London: British Medical Journal Publications, 1992.

Barlow, K. *Recognising Health*. London: The McCarrison Society, 1988.

Bateson, W. *Materials for the Study of Variation*. Cambridge: Cambridge University Press, 1894; reprinted by Johns Hopkins University Press, 1992.

Berger, S. and M. J. Kaever. *Dasycladales: an Illustrated Monograph of a Fascinating Algal Order*. Stuttgart: Thieme, 1992.

Berne Draft Resolution. *See* Altner.

Bohm, D., and B. J. Hiley. *The Undivided Universe: An Ontological Interpretation of Quantum Mechanics*. London: Routledge and Kegan Paul, 1993.

Brière, C., and B. C. Goodwin. "Geometry and Dynamics of Tip Morphogenesis in *Acetabularia*." *Journal of Theoretical Biology* 131 (1988): 461–475.

Cairns, J., J. Overbaugh, and S. Miller. "The Origin of Mutants." *Nature* 335 (1988): 142–145.

Cassirer, E. *Kant's Life and Thought*. London and New Haven: Yale University Press, 1981.

Church, A. H. *On the Relation of Phyllotaxis to Mechanical Laws*. London: Williams and Norgate, 1904.

Cole, B. J. "Is Animal Behaviour Chaotic? Evidence from the Activity of Ants." *Proceedings of the Royal Society London* (B) 244 (1991): 253–259.

Darwin, C. *On the Origin of Species*. (Reprint of the first edition of 1859.) London: Watts, 1950.

Davies, P., and J. Gribbin. *The Matter Myth*. Harmondsworth: Penguin, 1992.

Dawkins, R. *The Extended Phenotype*. Harlow: Longman, 1980.

———. *The Selfish Gene*. Oxford: Oxford University Press, 1976.

———. *The Blind Watchmaker*. Harlow: Longmans, 1986.

Delisi, C. "The Human Genome Project." *American Scientist* 76 (1988): 488–493.

Douady, S., and Y. Couder. "Phyllotaxis as a Physical Self-Organised Growth Process." *Physical Review Letters* 68 (1992): 2098–2101.

Fontana, W. "Algorithmic Chemistry," pp. 159–209 in *Artificial Life II: Santa*

Fe Institute Studies in the Sciences of Complexity, C. G. Langton, J. D. Farmer, S. Rasmussen, and C. Taylor (eds.), Vol. 10. Reading, Mass.: Addison-Wesley, 1992.

Fox, M. *Original Blessing.* Santa Fe, N.M.: Bear, 1982.

Franks, N. R., S. Bryant, R. Griffith, and L. Hemerik. "Synchronisation of the Behaviour within Nests of the Ant *Leptothorax acervorum.*" *Bulletin of Mathematical Biology* 52, (1990): 597–612.

Goldsmith, E. *The Way: An Ecological World View.* London: Rider, 1992.

Goodwin, B. C., and C. Briere. "A Mathematical Model of Cytoskeletal Dynamics and Morphogenesis in *Acetabularia,*" in *The Cytoskeleton of the Algae* (ed. D. Menzel), pp. 219–238. Boca Raton, Fla.: CRC Press, 1992.

Goodwin, B. C., and S. Pateromichelakis. "The Role of Electrical Fields, Ions, and the Cortex in the Morphogenesis of *Acetabularia.*" *Planta* 145 (1979): 427–435.

Goodwin, B. C., and L. E. H. Trainor. "Tip and Whorl Morphogenesis in *Acetabularia* by Calcium-Regulated Strain Fields." *Journal of Theoretical Biology* 117 (1985): 79–106.

Gould, S. J. *Wonderful Life. The Burgess Shale and the Nature of History.* New York and London: W. W. Norton, 1989.

———. "The Disparity of the Burgess Shale Arthropod Fauna and the Limits of Cladistic Analysis: Why We Must Strive to Quantify Morphospace." *Paleobiology* 17(4) (1991): 411–23.

Gould, S. J., and R. C. Lewontin. "The Spandrels of San Marco and the Panglossian Paradigm: a Critique of the Adaptationist Programme." *Proceedings of the Royal Society London* (B) 205 (1979): 581–98.

Green, P. B. "Inheritance of Pattern: Analysis from Phenotype to Gene." *American Zoologist* 27 (1987): 657–673.

———. "Shoot Morphogenesis, Vegetative through Floral, from a Biophysical Perspective. In *Plant Reproduction: From Floral Induction to Pollination.* (E. Lord and G. Bernier, eds.). *American Society for Plant Physiology Symposium Series,* Vol. 1 (1989): 58–75.

Harré, R., and E. H. Madden. *Causal Powers: A Theory of Natural Necessity.* Oxford: Basil Blackwell, 1975.

Harrison, L. G., and N. A. Hillier. "Quantitative Control of *Acetabularia* Morphogenesis by Extracellular Calcium: a Test of Kinetic Theory." *Journal of Theoretical Biology* 114 (1985): 177–192.

Harrison, L. G., K. T. Graham, and B. C. Lakowski. "Calcium Localization during *Acetabularia* Whorl Formation: Evidence Supporting a Two-Stage Hierarchical Mechanism." *Development* 104 (1988): 255–262.

Henderson, H. *Paradigms in Progress.* London: Adamantine Press, 1993.

Hodgkin, R. A. *Playing and Exploring: Education through the Discovery of Order.* London: Methuen, 1985.

Huizinga, J. *Homo Ludens: A Study of the Play Element in Culture.* London: Paladin, 1970.

Hull, D. "Historical Entities and Historical Narratives." In *Minds, Machines and Evolution* (ed. C. Hookway). Cambridge: Cambridge University Press, 1984.

Ingold, T. "An anthropologist looks at biology." (Curl Lecture, 1989). *Man* (NS) 25 (1990): 208–229.

Jaffe, L. F. "The Role of Ionic Currents in Establishing Developmental Pattern." *Philosophical Transactions Royal Society* (B) 295 (1981): 553–566.

Kauffman, S. A. *Origins of Order: Self-Organization and Selection in Evolution*. Oxford: Oxford University Press, 1992.

Kaye, H. *The Social Meaning of Modern Biology*. London and New Haven: Yale University Press, 1986.

Kortmulder, K. "The Congener: A Neglected Area in the Study of Behavior." *Acta Biotheoretica* 35 (1986): 39–67.

Langton, C. "Computation to the Edge of Chaos: Phase Transitions and Emergent Computation." *Physica* 42D (1990): 12–37.

———. "Life at the Edge of Chaos." In *Artificial Life II: Santa Fe Institute Studies in the Sciences of Complexity* (eds. C. G. Langton, J. D. Farmer, S. Rasmussen, and C. Taylor) Vol. 10. Reading, Mass.: Addison-Wesley, 1992.

Lewin, R. *Complexity: Life at the Edge of Chaos*. New York: Macmillan, 1992.

Lovelock, J. *A New Look at Life on Earth*. Oxford: Oxford University Press, 1979.

———. *Gaia: The Practical Science of Planetary Medicine*. London: Gaia Books Ltd, 1991.

Margulis, L. and D. Sagan. *Microcosmos*. London: Allen and Unwin, 1987.

Maturana, H. R., and F. J. Varela. *The Tree of Knowledge: The Biological Roots of Human Understanding*. Boston and London: New Science Library, Shambhala, 1987.

Mayr, E. *Toward a New Philosophy of Biology*. Cambridge, Mass.: Harvard University Press, 1988.

Miramontes, O., R. V. Solé, and B. C. Goodwin. "Collective Behaviour of Random-Activated Mobile Cellular Automata." *Physica D* 63 (1993): 145–160.

Mivart, J. St. George. *On the Genesis of Species*. London: Macmillan, 1871.

Norberg-Hodge, H. *Ancient Futures*. San Francisco, Calif.: Sierra Books, 1992.

Oster, G., and P. Alberch. "Evolution and Bifurcation of Developmental Programs." *Evolution* 36 (1982): 444–59.

Oster, G. F., and G. Odell. "The Mechanochemistry of Cytogels." *Physica* 12D (1984): 333–350.

Oster, G. F., J. D. Murray, and A. Harris. "Mechanical Aspects of Mesenchymal Morphogenesis." *Journal of Embryology and Experimental Morphology* 78 (1983): 83–125.

Oster, G. F., J. D. Murray, and P. Maini. "A Model for Chondrogenic Condensations in the Developing Limb: The Role of Extracellular Matrix and Cell Tractions." *Journal of Embryology and Experimental Morphology* 89 (1985): 93–112.

Oster, G. F., N. Shubin, J. D. Murray, and P. Alberch. "Evolution and Morphogenetic Rules: the Shape of the Vertebrate Limb in Ontogeny and Phylogeny." *Evolution* 42 (1988): 862–884.

Paley, W. *Natural Theology; or Evidences of the Existence and Attributes of the Deity, collected from the Appearances of Nature*. London: Printed for N. Faulder, 1802.

Rambler, M. B., L. Margulis, and R. Fester. *Global Ecology: Towards a Science of the Biosphere*. London: Academic Press, 1989.

Robertson, J. *Future Wealth: A New Economics for the Twenty-first Century*. New York: Bootstrap Press, 1990.

Sheldrake, R. *A New Science of Life*. London: Paladin, 1981.

Shiva, V. *Staying Alive: Women, Ecology and Survival in India*. London: Zed Books, 1988.

———. *Monocultures of the Mind*. Penang, Malaysia: Third World Network, 1993.

Shubin, N. H., and P. Alberch. "A Morphogenetic Approach to the Origin and Basic Organization of the Tetrapod Limb." *Evolutionary Biology* 20 (1986): 319–387.

Solé, R. V., O. Miramontes, and B. C. Goodwin. "Oscillations and Chaos in Ant Societies." *Journal of Theoretical Biology* 161 (1993): 343–357.

Sonneborn, T. M. "Gene Action in Development." *Proceedings of the Royal Society London* (B) 176 (1970): 347–366.

Spiegelman, S. "An *In Vitro* Analysis of a Replicating Molecule." *American Scientist* 55 (1967): 221–264.

Stewart, I. *Does God Play Dice?* Harmondsworth: Penguin, 1989.

Stewart, I. and M. Golubitsky. *Fearful Symmetry: Is God a Geometer?* Oxford: Blackwell, 1993.

Swinney, H., J-C. Roux, and R. Simoyi. *Physica* 7D (1983): 3–15.

Thompson, D. W. *On Growth and Form*. Cambridge: Cambridge University Press, 1917.

Turing, A. M. "The Chemical Basis of Morphogenesis." *Philosophical Transactions of the Royal Society* (B) 237 (1952): 37–72.

Varela, F. J., and E. Thompson. "Color Vision: A Case Study in the Foundation of Cognitive Science." *Revue de synthèse* IV, 5, Nos. 1–2 (1990): 129–138.

Waldrop, M. M. *Complexity: The Emerging Science at the Edge of Order and Chaos*. New York: Simon and Schuster, 1992.

Wemelsfelder, F. *Animal Boredom*. Utrecht: Elinkwijk BV, 1993.

Whitehead, A. N. *The Concept of Nature*. Cambridge: Cambridge University Press, 1920, 1971.

Williamson, G. S., and I. H. Pearse. *The Case for Action*. London: Faber, 1931.

Winfree, A. T. *When Time Breaks Down*. Princeton, N.J.: Princeton University Press, 1987.

Winnicott, D. W. *Playing and Reality*. London: Tavistock, 1971.

Wolfram, S. *Theory and Applications of Cellular Automata*. Singapore: World Scientific, 1986.

Wrench G. T. *The Wheel of Health*. New York: Schocken, 1972.

de Beer, G. *Homology: An Unsolved Problem*. Oxford: Oxford University Press, 1971.

Dobzhansky, Th. "Nothing in Biology Makes Sense Except in the Light of Evolution." *American Biology Teacher*, March 1973: 125–129.

Emmet, D. *The Effectiveness of Causes*. London: Macmillan, 1984.

———. *The Passage of Nature*. London: Macmillan, 1992.

Frankel, J. *Pattern Formation*. Oxford: Oxford University Press, 1989.

Goodwin, B. C. "Development and Evolution." *Journal of Theoretical Biology* 97 (1982): 43–55.

———. "Structuralism in Biology." *Science Progress* (Oxford) 74 (1990): 227–244.

———. *Development*. London: Hodder and Stoughton and The Open University, 1991.

———. "Development as a Robust Natural Process." In *Thinking About Biology* (W. D. Stein and F. Varela, eds.). Reading, Mass.: Addison-Wesley, 1993.

———. "Towards a Science of Qualities." In *The Metaphysical Foundations of Modern Science* (ed. J. C. Clark). Sausalito, Calif.: Institute of Noetic Sciences (in press), 1994.

Goodwin, B. C., S. A. Kauffman, and J. D. Murray. "Is Morphogenesis an Intrinsically Robust Process? *Journal of Theoretical Biology* 163 (1993): 135–144.

Goodwin, B. C., A. Sibatani, and G. C. Webster (eds.). *Dynamic Structures in Biology*. Edinburgh: Edinburgh University Press, 1989.

Goodwin, B. C., J. C. Skelton, and S. M. Kirk-Bell. "Control of Regeneration and Morphogenesis by Divalent Cations in *Acetabularia mediterranea*." *Planta* 157 (1983): 1–7.

Goodwin, B. C., and P. T. Saunders (eds.). *Theoretical Biology: Epigenetic and Evolutionary Order from Complex Systems*. Edinburgh: Edinburgh University Press, 1989.

Gould, S. J. *Ontogeny and Phylogeny*. Cambridge, Mass.: Belknap Press, 1977.

Hall, B. G. "Adaptive Evolution That Requires Multiple Spontaneous Mutations." *Genetics* 120, (1988): 887–897.

Hall, B. K. *Evolutionary Developmental Biology*. London: Chapman and Hall, 1991.

Jarvik, E. *Basic Structure and Evolution of Vertebrates*. Vol. 2. London: Academic Press, 1980.

Murray, J. D. *Mathematical Biology*. Berlin: Springer Verlag, 1989.

Nicolis, G., and I. Prigogine. *Exploring Complexity*. New York: W. H. Freeman, 1989.

Oosawa, F., M. Kasai, S. Hatano, and S. Asakura. "Polymerisation of Actin and Flagellin." In *Principles of Biomolecular Organisation* (ed. G. E. W. Wolstenholme and M. O'Connor). Boston, Mass.: Little, Brown, 1966.

O'Shea, P., B. Goodwin, and I. Ridge. "A Vibrating Electrode Analysis of Extracellular Ion Currents in *Acetabularia acetabulum*." *Journal of Cell Science* 97 (1990): 505–508.

Wolpert, L. "Positional Information and the Spatial Pattern of Cellular Differentiation." *Journal of Theoretical Biology* 25 (1969): 1–47.

———. *The Triumph of the Embryo*. Oxford: Oxford University Press, 1991.